Lean Compendium

AF768817

Bruno G. Rüttimann

Lean Compendium

Introduction to Modern Manufacturing Theory

Springer

Bruno G. Rüttimann
D-MAVT - IWF
inspire AG/ETH Zürich
Zürich, Switzerland

ISBN 978-3-319-58600-7 ISBN 978-3-319-58601-4 (eBook)
DOI 10.1007/978-3-319-58601-4

Library of Congress Control Number: 2017942753

© Springer International Publishing AG 2018
This work is subject to copyright. All rights are reserved by the Publisher, whether the whole or part of the material is concerned, specifically the rights of translation, reprinting, reuse of illustrations, recitation, broadcasting, reproduction on microfilms or in any other physical way, and transmission or information storage and retrieval, electronic adaptation, computer software, or by similar or dissimilar methodology now known or hereafter developed.
The use of general descriptive names, registered names, trademarks, service marks, etc. in this publication does not imply, even in the absence of a specific statement, that such names are exempt from the relevant protective laws and regulations and therefore free for general use.
The publisher, the authors and the editors are safe to assume that the advice and information in this book are believed to be true and accurate at the date of publication. Neither the publisher nor the authors or the editors give a warranty, express or implied, with respect to the material contained herein or for any errors or omissions that may have been made. The publisher remains neutral with regard to jurisdictional claims in published maps and institutional affiliations.

Printed on acid-free paper

This Springer imprint is published by Springer Nature
The registered company is Springer International Publishing AG
The registered company address is: Gewerbestrasse 11, 6330 Cham, Switzerland

To my son Marco and to the memory of my parents.
Hic liber vade mecum est ubi sapientia gaudet adiuvare usu.

Foreword

Lean Compendium: Introduction to Modern Manufacturing Theory

Fabricare necesse est—but how?

The industrial production has witnessed within the last decade a drastic change in the reputation in public and mostly political view. Ten years before, the general opinion on production in highly developed countries was that industrial value adding is old-fashioned and needs to decay, while future-oriented industrial nations develop in the direction of a service society and aim at a share of industrial value creation of 10% of BIP and less. Today especially those countries are in an unfortunate situation and look desirously onto those economies formerly termed as old-fashioned that have kept their industrial value creation up at over 20%. It was clearly recognized that only non-value-adding businesses do not sustain a national economy. Where the limits of stability are is not really known, but all the reshoring initiatives today in highly developed and thus high-wage countries demonstrate impressively the societal value of industrial production, despite the fact that in the foreground the motivation is to keep or bring back the most recent and fashionable technological developments, the hypes of manufacturing like additive manufacturing, industrie 4.0 and formerly nanotechnologies. Following up the growth of the tertiary economic sector, it must be kept in mind that a good share of this is due to outsourcing of services from producing industries, and those services would also vanish by reducing the industrial value creation. The local industry is the best customer for service organizations.

It shall not be denied that quite some of the industrial value creation eroded away and became outsourced to low-wage countries, which are fierce competitors to industrial value creation in highly developed countries. Production was regarded as an easy task, so easy that it could be managed in underdeveloped areas of the world. Production was in comparison to old times, where the fuming factory chimneys were the symbols of status and wealth, no longer looked at as being worthwhile to sustain, which ended in disregard, divestment and thus old-fashioned equipment and organizations. This is then the point, where outsourcing becomes inevitable. But the most important error was to not realize that efficient production is not a simple task, just an annex to the ingenious product creation and sales.

It is the complexity of production, lack of strict scientific approaches, number of influencing variables, restrictions, interrelations, knowledge and experience including aspects of finances, resources, people and technologies which makes seeking for optimal solutions in production so difficult and leads to industrial managers hoisting the white flag of surrender, which is called outsourcing. It seemed so easy, just to get rid of the management-attention-swallowing burden of manufacturing and deport it to places, where due to low wages almost every however stupid organization could survive.

Since the task today is facing the huge competition of globalization and challenges of ecologic sustainability to achieve more with less resources, the resource-based manufacturing needs to be changed to knowledge-based manufacturing. And against plagiarism such knowledge-based manufacturing is the best way to protect intellectual properties. While the innovation in product development needs to be presented in the market and thus to the Argus eyes of plagiarism, the innovation in processes, which means striving for technological as well as organizational excellence, can be kept non-disclosed and guarantees a threefold time span competitive advantage over the product innovation. The transition towards knowledge-based manufacturing requires highest skills and excellent leadership from the production management and is at the same time the chance to answer the question how to keep industrial production in the high-wage countries with their well-trained and skilled workforce. All the former fundamental industrial changes also termed as industrial revolutions aimed at exploiting technological innovations for efficiency in production, not really taking into account that also the organization needs to follow. And also the most recent technology change, introduction of Internet technologies, called Industry 4.0, finds industry orientationless especially on how to apply this technology for business excellence, despite some quite striking examples of benefit already demonstrated in economy and to a lesser extent even in production.

But towards operational excellence, the great change, if not revolution, took place with Toyota Production System (TPS) or synchronized production system (SPS) after the Second World War. Here a solution for operational excellence was presented completely independent of technologies and thus ever valid. For the Western world, this came out of the nothing, outperforming traditional production organizations. It is based on simple principles such as removing the nine kinds of muda, pulled material flow and permanent improvements. However, outside of automotive industry, this or similar systems have not fully been received, which might be due to disregard of production as such or still not realizing the great chance offered by a disruptive change of production philosophy.

Now, this lean compendium written by a real expert, combining scientific insight with practical experience, offers the great chance to enter into the fascinating modern art of production of the best performing production systems in the world. It combines logical and enlightening arguments and explanations with clear proposals on how to set up such systems in production of whatever size and product. It succeeds in presenting a comprehensive basis for a theory of production, on which a systematic optimization can be built upon, thus getting rid of beliefs and fashions in favour of a data and reality-based assessment and improvement of production. It serves as a guideline on how to salvage production industry in high-wage countries serving as railing as well as life buoy in one.

This book, although for sure not complete and comprehensive, is a good base to understand what the TPS, the today's best performing production system, consists of, which helps productioneers to improve present manufacturing systems radically. This book is for sure an enrichment of present production literature for students as well as experienced engineers and managers and all those who want to know how a modern and competitive production system is really functioning.

IWF Institute of Machine Tools and Manufacturing
ETH Zürich, Swiss Federal Institute of Technology
Zurich
Switzerland
12 April 2017

Prof. Dr.-Ing. Konrad Wegener

Prologue

Lean Manufacturing (LM) is often seen as a hands-on practitioner-driven approach to implement the Toyota Production System (TPS). This is not wrong—introducing Lean needs a lot of perseverance to succeed. Taiichi Ohno once said: "The challenge is to develop a learning organization that will find ways to reduce the number of Kanbans". Indeed, the importance and criticality of the practical human dimension of improving a Lean system has not to be underestimated. Continuous improvement is the winning management philosophy how to run successfully a company from the operational point of view to face and to adapt to new challenges. Even so, the implicit theoretical dimension of Lean is often neglected and seldom seen comprehensively formalized as if it was a given fact. To the contrary: besides Kaizen and Muda elimination, at the base of the TPS is a solid theoretic framework which rarely is talked about because of not being formalized or, if ever, limited to a trivial didactic simulation comparing a single piece flow (SPF) versus a batch and queue (B&Q) manufacturing modus explaining Little's Law.

Therefore, in this compendium—being in fact a "vademecum" which should accompany every Lean expert—we are entering into the dimension beyond Muda and Kaizen; we will consolidate the underlying manufacturing theory of Lean with regard to performance. We will structure and formalize the existing fragmented Lean theory framework of manufacturing concepts into generalized practitioner-conforming production laws, putting together a systematic and integrated set of generally valid theorems and corollaries which govern manufacturing, helping to form ex post the base of LM systems theory. In a nutshell, we will give a solid structure to Lean beyond the fuzzy TPS philosophy. On the other hand, we will not repeat statistics or queuing theory, well described in many academic textbooks, nor will we talk about the Lean management tools as well as shopfloor concepts and Kaizen practices of a learning organization that are needed to run and improve the TPS; other books already exist on these topics. We want to enter the dimension between practical description and academic theory. We will use the strict minimal necessary math combined with a comprehensible language for engineers and practitioners as well as students of mechanical engineering curriculums to understand the underlying theory of Lean, i.e. the omnipotent governing technical principles, which characterizes TPS. Thus, the intention is not to write another book about Lean but to complement the existing literature. Nevertheless, this compendium represents the copestone to bridge the gap between basic description of Lean systems and its related analytical manufacturing theory.

Dr.-Ing. Bruno G. Rüttimann

Acknowledgements

My very special thanks go to Dr. Martin Stöckli, manager of the inspire academy and COO of inspire AG, a technology transfer institute of ETH Zürich, the Swiss Federal Institute of Technology, promoted by Prof Konrad Wegener, for having given precious advice for this compendium.

Contents

Chapter 1
Introduction

The TPS has become the reference of modern high performance manufacturing systems. It has been spread and adopted in Europe under the American label of Lean Manufacturing (LM) also among other industries than the automotive because of its superior performance. Apart of the Kaizen-based continuous improvement management philosophy, the underlying TPS theory bases on a Just-in-time (JIT) type manufacturing approach. This manufacturing approach bases on "flow on pull" with Heijunka-pitch scheduling, i.e. self-controlled mixed-product cells, which performance is by far higher than those of traditional computer-controlled and optimized MRP 2-type (manufacturing resource planning) or ERP-type (enterprise resource planning) systems relying mainly on push "batch & queue" (B&Q) manufacturing. Indeed, Lean stands in contradiction to the Western "high performance" thought of B&Q manufacturing of large batches to minimize setup downtimes and reducing cost per piece, exploiting equipment output, keeping blue collar workers busy and hurrying, i.e. showing an apparent high productivity. But the high level of busyness may contain a lot of non-necessarily needed activities, such as searching, bringing, handling, piling, waiting, so-called non-value add activities or inefficiencies which the Japanese call Muda (waste). Fujio Cho has defined waste as "anything other than the minimum amount of equipment, materials, parts, space, and worker's time, which are absolutely essential to add value to the product". Instead of a hurrying activism, the Japanese prefer not a calm but a waste-less sequence of activities at a constant pace resulting at the end of the day in a higher efficiency and efficacy.

Another difference is the concept of built-in quality of the TPS, every employee contributing to guarantee error-free products not necessitating final inspection. Exactly the final check has been typical of Western production systems to guarantee quality, scrapping defective products at the end of the line bearing the highest value content. The Japanese even foster the culture to allow blue collar workers to stop the assembly transfer line in case they discover a non-conformity. Until not many years ago, such a contentious behavior in a Western automotive plant would have lead to the immediate consequence of being fired. At the base to implement such an

© Springer International Publishing AG 2018

B.G. Rüttimann, *Lean Compendium*, DOI 10.1007/978-3-319-58601-4_1

error-free culture stands the SPF, allowing to discover defects and to act immediately solving the problem. SPF and Jidoka therefore go hand in hand to guarantee a JIT zero mistake culture which allows on-time-delivery (OTD). The cultural change needed is drastic but the resulting performance improvement showed the benefit [1].

During the last two decades also the industry logic changed profoundly. Globalization revolutionized the way of thinking. Globalization brings not only more opportunities through a larger world market, it entails also more competition and therefore a threat. Different globalization forms are observable bearing different underlying business types with different rationales [2]. Understanding the changing rationale by understanding the development of value-add chains is decisive to survive the competitive challenge. These changes have transformed vertically integrated value-chain business models into horizontal networks exploiting rather economies of scope than scale [3]. The industry logic and consequent operational rational has changed. Indeed, in the meanwhile the automotive industry logic has mutated from a value-add per car to a value-add per hour logic [4]. Indeed, to increase brand recognition with multiple beneficial effects forces management to increase throughput to have more cars circulating on streets. Western mature economies will face increased competition and will forcedly need to change behavior regarding marketing mix and production. Innovation is seen as a "deus ex machina" paradigm; this is not wrong but the companies need also to produce these new products with acceptable cost. Otherwise the epitaph on the tombstone of Western high cost industries will read: "He was a good inventor but a bad producer" [5].

To have success in today's intensive competitive environment not only the product has to be the best; to provide a unique selling proposition, also the price has to be aligned as well as the ancillary boundary conditions, such as immediate availability and service have to be observed. The SPQR model summarizes these requirements [6]. Here, SPQR does not stand for "Senatus PopolusQue Romanus" but for the today's necessary minimal competitive cardinal variables to be satisfied: speed (i.e. process lead time PLT), punctuality (i.e. on-time delivery OTD), quality (i.e. Z-level), as well as a minimum of attractive return (profit) for the investor. Return has to be part of the equation because only a profitable company allows the system to be viable. These system variables are interacting with the stakeholder variables customer (voice of customer VOC and satisfaction), employee (voice of employee VOE, appreciation, empowerment, and satisfaction), shareholder (return on investment ROI and satisfaction) and give origin to a positive (amplifying positively or negatively according to the input) feedback governed system dynamics. Moreover, the model shows clearly, that the Western investor-centric philosophy around shareholder satisfaction is short-term oriented and does not represent a long-term viable model. This also reflects the sometimes encountered mess between vision and mission statement of governing company statements. Indeed, it is the customer stakeholder-centric model which is the only long-term viable winning solution, exactly what the Toyota philosophy of acting represents. All stakeholders have to contribute to value-generation but also to receive back their

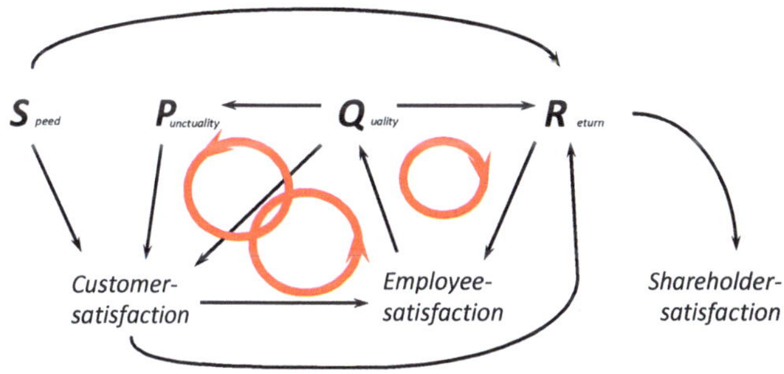

Fig. 1.1 SPQR model showing systemic effects between the cardinal objectives [from 6]

compensation according to their personal value system in order to be satisfied to give their max. Further, this SPQR model shows the central place of JIT Lean, but also Six Sigma, within the whole operations performance. It shows the crucial importance of the employee on quality and the other variables. The true assets of a company are the employees and customers—not the machines, which become obsolete and have to be amortized, and for sure not the inventory—allowing the solution of the "money-for-value" equation. The basic SPQR requirements can be considered to be time-invariant and constitute a sort of a minimal axiomatic multi-objective system which has to be observed in any case to be successful in business (Fig. 1.1).

Under the Lean label, the TPS has been described and divulged in several books, e.g. [7–10]. Nevertheless, they are stressing far more the philosophy of the TPS and the Kaizen-based shopfloor continuous improvement culture of a learning organization as well as describing the believed omnipotent tools rather than the implicit theoretic governing laws of the TPS. The theory framework is widely ignored or neglected [11, 12]. However, different than in Anglo-Saxon countries, in Europe, and especially in Switzerland, LM is only making a reluctant appearance at university level [13] showing the missing of a driving high performance-demanding automotive industry. We will not investigate the further root causes regarding the scarce presence of Lean courses. However, we will put the attention to bridge the lack of divulgation of the underlying theoretic and implicit manufacturing framework of Lean. In order to educate state of the art production managers and engineers, the production theory of TPS has to be taught, opening the "academic dimension" of Lean not only for students but also for already field-proven practitioners being in responsible positions, often consisting of engineers with now obsolete knowledge. It is not the responsibility of industry to educate production managers with Lean Sensei or Lean Six Sigma Black Belt courses aligning them with "state-of-the-art" knowledge.

This compendium intends to bridge application and theory by following a Cartesian-type logic of reasoning to model Lean, although it does not pretend to be comprehensive in all topics. It begins by presenting a new cognitive-oriented, systemic-based Lean model, i.e. describing the systems-based functioning of Lean.

It is modeled by a new, structured mono-pillar representation of the TPS, where JIT and Jidoka are not separately shown in two pillars, as usual [11, 14]. Indeed, the two concepts are not only complementary but synergic. This book explains the basic concepts of queuing and WIP formation by a comparative evaluation between the two main manufacturing modi, i.e. single piece flow (SPF) as well as batch and queue (B&Q), defining unequivocally the concept of bottleneck. It displays the underlying production laws translated into manufacturing language for proven and novel production engineers such as the necessary and sufficient condition of the basic customer requirement for an on-time-delivery (OTD), explaining the introductory concepts of a manufacturing system (Chap. 3). It introduces and defines the concept of manufacturing lead time (MLT) and the difference to the better known process lead time (PLT) and how they are linked. It analyzes the relative performance of alternative manufacturing modi with balanced and unbalanced cycle times characteristics. Chapter 4 analyzes comparative performances of manufacturing systems and defines most of the general fundamental laws leading to general manufacturing theorems, corollaries and lemmas. We will intend hereafter corollary as an auxiliary statement to the related theorem and lemma as a derived statement of more practical application. Further, we will list several types of principles, where principles describe technical ways how to implement a production related aspect. Whereas theorems and corollaries correspond to production laws, most principles can be chosen deliberately, but there are principles which suit better or less a certain transformation environment. Notice, different principles may have different performance in terms of speed and punctuality. Simulations will show the influence of variance on lead time. We will space from mono-product manufacturing (Chap. 4), to mixed-model cells (Chap. 5), and complex manufacturing systems of interlinked cells (Chaps. 6 and 7). Chapter 8 is only for completeness; it introduces into how to deploy and maintain a Lean system and shows some differences between Lean and the new envisaged Industry 4.0 revolution.

The focus of this compendium is to give an introductory but solid, at the same time concise fundament about manufacturing theory leading naturally to the TPS model. It provides the mathematical evidence of the superior performance of the TPS. To do so, simplifications have to be made; indeed, we will assume no quality issues, no breakdowns, no lack of material, i.e. ideal manufacturing conditions, being breakdowns and quality issues rather a consequence of poor management how to run a plant, which, by the way, Toyota has put a lot of effort to solve. Further, quality and how to generate quality is essential within Toyota, having led to the concept of autonomation (Jidoka), i.e. automatic but at the same time foolproof equipment, allowing the operator to manage several workstations at the same time (Chaku chaku). Although the aspect how to generate quality is central to the TPS—indeed, SPF allows to generate quality by bringing defects to the surface letting inefficiencies (Muda) emerge—we are interested here in the working mechanism of the manufacturing system, becoming product quality and equipment uptime, such as other requirements, each one just an aspect of the JIT approach. This might be controversial, being Jidoka and total productive maintenance (TPM) at the base of Lean. In this book however, the focus will be put primarily on the formalized governing JIT functioning of the TPS. This compendium has to be seen as a first essay, for sure improvable and extendable, to give not only students a simplified but quite comprehensive mathematical view about Lean, educating modern production managers.

References and Selected Readings

1. Womack, J.P., Jones, D.T., Roos, D.: The Machine that Changed the World. Free Press, New York (1990)
2. Rüttimann, B.: Modeling Economic Globalization—a Post-Neoclassic View on Trade and Competition. MV-Wissenschaft, Münster (2007)
3. Rüttimann B, 2011, The Globalizing Aluminium Industry—Lean Supply Chains Will Shape a New Industry Logic, Part 2, ALUMINIUM 87, 9, Giesel Verlag, Hannover
4. Hagen, H., Rüttimann, B.: The Automotive Market—the New Challenge for the Aluminium Industry, ALUMINIUM 80, 3/4. Giesel Verlag, Hannover (2004)
5. Rüttimann B.: Globalization and the strategic challenge. Presentation held at: Swiss-Swedish innovation initiative, EMPA Dübendorf, September 25 (2013)
6. Rüttimann, B.: The Central Importance of Quality, ALUMINIUM 77, 7/8. Giesel Verlag, Hannover (2001)
7. Suzaki, K.: The New Manufacturing Challenge. Free Press, New York (1987)
8. Ohno, T.: Toyota Production System—Beyond Large Scale Production. Productivity Press, New York (1988)
9. Womack, J.P., Jones, D.T.: Lean Thinking. Free Press, New York (2003)
10. Liker, J.K.: The Toyota Way, 14 Management Principles from the World's Greatest Manufacturer. McGraw-Hill, New York (2004)
11. Rüttimann, B., Stöckli, M.: Going beyond triviality: The Toyota production system—lean manufacturing beyond Muda and Kaizen. J. Serv. Sci. Manag. **9**, 140–149 (2016)
12. Rüttimann, B.G.: Discourse about linear programming and lean manufacturing: Two different approaches with a similar, converging rational. J. Serv. Sci. Manag. **8**, 85–91 (2015)
13. Rüttimann, B., Wegener, K.: Einführung in die Methoden von Lean Manufacturing und Six Sigma Quality Management, ETH Tools-IV Kurs, Lecturing notes HS2014, D-MAVT (2014)
14. Rüttimann, B.: Von Lean zu Industrie 4.0—eine Evolution? Von einer visionären Idee zum realen Verständnis, Presentation held at: Fertigungstechnisches Kolloquium, ETH Zürich, November 26, 2015 (2015)

Chapter 2
Modeling of Production Systems

In the following sections we will introduce the reader to the complexity of production systems. We will revisit the emblematic "two-pillar" temple model of the TPS giving to the systemic characteristics of the TPS also a suitable systemic representation by introducing an integrated "mono-pillar" model. This new Lean-systemic TPS model will be "le fil rouge" across this compendium to describe production theory of Lean. Furthermore, based on physical analogies we will enter into the basic concept of flow and define a thermodynamics-derived system of Lean Governing Principles.

2.1 Optimization of a Complex System

Different from exact natural sciences such as physics, chemistry, or mathematics, production theory is rarely developed in a purely analytical way, despite laws are governing the production logic. This might originate from the fact that production, i.e. the transformation of inputs into outputs, apparently is not the same such as physics described by mathematical transfer functions $y = f(x)$ with a deterministic solution. However, production is a multidimensional science of

- application of production-related math
- use and allocation of various limited resources
- respect of economic requirements
- within a non-deterministic environment, allowing different possible but also non-optimal solutions.

The final aim of production is to create value for society complying with a long-term sustainable company mission. Within a specified timeframe with limited resources, the economic requirements are to transform input factors such as different raw materials into output objects such as intermediate components or usable

© Springer International Publishing AG 2018

B.G. Rüttimann, *Lean Compendium*, DOI 10.1007/978-3-319-58601-4_2

products. The transformation has to be performed in an optimal way; economists use to say

- maximize output with a given input, or
- minimize inputs to obtain a specified output.

For a production system, usually the latter applies. Different than physics and chemistry, production is a complex interdisciplinary science. The complexity is given by the characteristics of the multiple subsystems involved as well as the multiple degrees of freedom to realize the transformation. The main subsystems are:

- available process resources x_p, such as machines, workforce, and time with limited capacity (in a given timeframe)
- balance on hand of various input resources x_i, such as raw materials and components
- manufacturing laws of the applied transformation process governing the transfer function f(x)
- two, initially distinct, multi-objective functions of demand (customer) z_1 and supply (producer) z_2 integrated finally by following the SPQR axioms.

Different from most of operations research techniques, which optimize not only but mainly static problems (see e.g. Sect. 3.3 Queuing theory and WIP formation), the complexity of the system is further augmented by the temporal dynamics of succeeding random customer orders showing also high product-mix variability. To manage the delivery requirements within an evolving not-static context, time horizon of planning is fractioned and often solved with weekly scheduled production campaigns. We will see that Lean is skipping this rigid campaign model; by the way, which also flexible Industry 4.0 systems will try to do, nevertheless following a different approach (see Sect. 8.3 Lean and the fourth industrial revolution). We will not enter here into theoretic, complex modeling of multi-level systems with operations research techniques; however, we will see how Lean solves the complexity problem of production. In the following, we will define and intend production as:

- the optimization of a constraint system
- with the objective to transform input factors into products (physical transformation)
- complying to customer requirements such as OTD (VOC)
- having limited process resources available (capacities)
- applying an appropriate allocation, i.e. scheduling of resources (optimal solution)
- by following the economic rational of minimizing waste of input and resources (ROI).

This definition shows how complex it is to manage a production system.

2.2 Reconsidering the TPS: The Systemic Lean Model[1]

After WW2, over more than three decades, Toyota implemented step by step a comprehensive proprietary manufacturing system which in the 1980s became known as the TPS. This TPS has been growing organically meaning that it has been conceived continuously based on common practical sense and integrating acquired experience, questioning present Western manufacturing systems based on B&Q principle. One of the first books describing the system was written by Taiichi Ohno, considered the father of the TPS [2]. According to Liker [3], the Toyota house is attributed to Taiichi Ohno's disciple Fujio Cho, who developed the model to teach the TPS to suppliers. The model is of cognitive type, structuring the components of the TPS. It shows the foundation on what it bases, the two reinterpreted novelties of flow and quality, which can considered to be the two sustaining pillars of the TPS, as well as the team-based Kaizen to reduce Muda (Fig. 2.1). For further description of this classical TPS model and the related Lean tools, we refer to the existing literature, e.g. [3, 4].

Of this classical two-pillar temple model, a large number of different interpretations, of more simple or more complex TPS representations exists. Nevertheless, they all feature the same lack, such as describing the TPS as a list of topics, which has been leading several Western companies to adopt just some of the indicated tools interpreting the TPS like a tool-box. This might have been the consequence of overstressing the ultimate mantra of waste reduction, losing the comprehensive

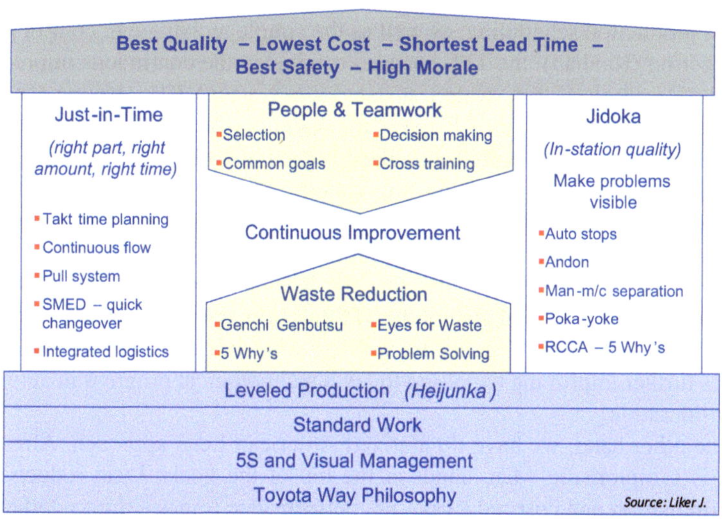

Fig. 2.1 The widespread representation of the TPS is the classical two-pillar temple house model, as depicted by e.g. [3]; note, TPM is missing

[1]The main part of this section has been taken as excerpt from [1].

concept which stands behind the TPS. Indeed, despite showing the TPS model a structure, it seems to be interpreted as an amorphic structure. This sounds to be a contradiction because amorph means to have no structure. What I mean is, the missing of logic interconnections and implications of each subsystem leading to a solid systemic structure.

The original Lean philosophy compared to the original TPS is simplified and usually "sold" as waste reduction. This is not wrong per se, but the aim of TPS is much higher, with waste reduction being quasi a natural by-product if the TPS is well applied. Indeed, Toyota developed exactly the necessary concepts and techniques to limit, or better to avoid, waste production having scarce availability of resources. A synergic system of techniques have been put together around SPF (which, nota bene, is not a Toyota invention, but based on Taylorism and applied first in Ford's T-model production) to allow a flawless quality-oriented operation of a SPF without waste, as well as Heijunka-box leveled pitch to limit WIP (work in process, which is also considered as waste and which delays PLT) and increasing flexibility. Such as the TPS acronym suggests, the emphasis is put on the production system. It is a new way how to produce, how to maximize the output of assembly process type of operations by speeding up PLT, integrated by in-station guaranteed product quality. The TPS has not been conceived by applying manufacturing theory (what we try to do with this book), but by attentive observation and evaluation how to best eliminate any waste and optimize process performance (learning by doing, i.e. observing and improving). Interesting is, that Toyota does not eliminate Muda per se, but via elimination of Mura, i.e. smoothening unevenness. Apart of the underlying tools (SMED single minute exchange of die or Heijunka-pitch) to create a smooth production scheduling, as well as the simple technique to control production triggering (Kanban), the TPS has also originated the continuous improvement approach (Deming's PDCA cycle translated into Kaizen). The striving for perfection by using the "hidden" knowledge of the operators at the shopfloor level, where production takes place (Gemba), has been copied already very early by Western companies, creating the suggestion box system. This was a first timid attempt to implement the continuous improvement process, however far away from how it has been intended by the Japanese Kaizen approach; by the way, in Switzerland it is still believed by 10% of "professionals" that the Kaizen approach corresponds to the suggestion-box [5]. The final goal of the TPS has been the wasteless JIT production. At the end, it took three decades to develop what is called TPS today, and the system is further improving by taking today's technological progress in automation into consideration.

On the other hand, we have the derived American Lean approach. Already the naming is symptomatic what stands at the top of the goals: Lean reflects speed, waste elimination and cost reduction, i.e. performance translated into dollars. This is the straight forward oriented approach of Western enterprises to catch-up. The usually most taught Lean concepts are mainly all about VSM (Value Stream Mapping) and Muda identification and elimination as well as Womack's Lean Transformation approach [4]. In addition, a strong tool-based belief is at the core, which often deviates from the real origin of the problem itself. This is a different approach than the original TPS. This is not surprising; indeed, the TPS is an organically grown production system, a production philosophy, whereas Womack's Lean Thinking [4] is the propagation of a "recipe" to catch-up fast in order to become again competitive. Although Lean is often superficially used as synonym to TPS, the rationale behind and the approach is clearly different but not the goal.

It has also to be explicitly stated that the TPS has been developed to optimally match the assembly type of production, but this does not mean, that it is not applicable to other types of manufacturing systems, as the Japanese shipyards already showed in the eighties. Nevertheless, in non-assembly industries, the Toyota production theory and techniques are reluctantly implemented, because resulting sometimes difficult to interpret the concepts and therefore how the tool has to be adapted to the different process characteristics. The consequence is to use only a part of the Lean tool set limiting the exploitation of the real improvement potential. The limited use of the tools might also stem from the classic two-pillar temple house representation of the TPS (Fig. 2.1) which might mislead to pick a few suitable tools just as needed.

To highlight the synergic interaction of the systemic TPS elements, it is advisable to teach students the Lean approach with an integrated presentation of the JIT and Jidoka concepts within a mono-pillar model as shown in Fig. 2.2 [6]. Indeed, being the final aim to have "the right product with the right quality" the Jidoka based in-line or in-station quality should not be shown separate from the JIT flow

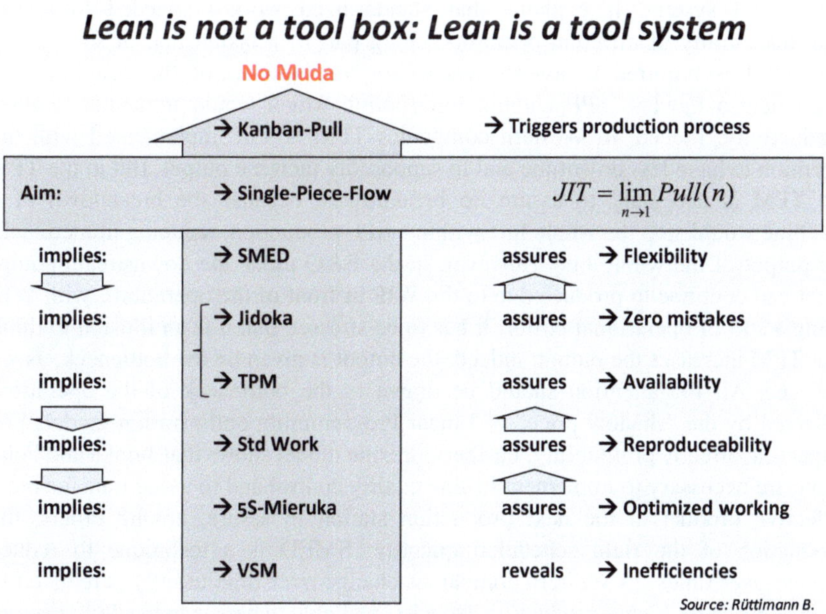

Fig. 2.2 The systemic mono-pillar Lean model [from 6] shows the basic ideas behind the synergic tool system

pillar, as displayed by the classic two-pillar TPS temple model of Fig. 2.1, built-in quality becoming one element among others to implement a flawless SPF. If ever a "multi-pillar" representation may be used, the parallel concept to Jidoka (manufacturing quality parts) should also embrace TPM (Total Productive Maintenance) assuring the correct functioning of the equipment, with TPM being as much important as quality to allow a flawless SPF. Such as the various techniques of Jidoka (Poka Yoke, Andon, stopping line culture, etc.) also TPM techniques (maintenance prevention, preventive maintenance, predictive maintenance, autonomous maintenance) have to be put into place to guarantee the full operability of the line at any time. Be aware, in the case of a breakdown of a machine you do not have a WIP decoupling buffer as operational reserve in between of the operations to continue production; you must guarantee the problem free functioning of the equipment—that is the reason to have the overall equipment effectiveness OEE indicator in place.

This mono-pillar representation shows the cascaded requirements to implement a flawless SPF. This model is a first attempt to show the system's interactions in a simplified way between the main Lean tools. It points at the intrinsic aim to have a SPF in order to gain speed for reducing process lead time and to increase productivity. It shows also that Kanban stocks are not the aim, despite it is sometimes understood so. The model of Fig. 2.2 reveals a quite different aspect of the TPS than the two-pillar temple model of Fig. 2.1; namely the real intrinsic nature of the theory of Lean regarding the Lean tools (neglecting for simplicity the Kaizen aspect of Lean, i.e. continuous improvement). It shows clearly that Lean is not a toolbox, but a tool system. It explains that standardized work is needed to assure reproduceability of different operators, being part of a takted line. It shows also that TPM is required to assure availability, i.e. uptime, of the equipment to implement a flawless SPF without interruption which would immediately limit productivity. Indeed, in Western companies TPM is still implemented with the intention to have less downtime and to supposedly increase output. But in the TPS, the TPM is necessary to assure no breakdowns, because the breakdown of a machine would stop the whole line within a SPF production, reducing immediately the output of the whole line. However, in the B&Q mode the downstream equipment can continue to produce due to the WIP in front of the operations, with WIP being a sort of operational buffer. It has to be stressed that it is an illusion to think that TPM increases the output; indeed, the output is given by the bottleneck, as we will see. All the attention should be drawn to the bottleneck of the operation, reflected by the "shadow price" of Linear Programming optimization models [7], impacting directly profitability. Furthermore, the model shows that Jidoka and Poka Yoke are necessary to implement in-line quality control and to avoid transferring a defective product to the next production station to assure, among others, the production of the right scheduled quantity. SMED is a technique to reduce change-over times. In Western companies change-over time usually is reduced to have supposedly higher production capacity available, whereas in the TPS change-over time is reduced to allow mixed-product cellular manufacturing for a Heijunka box pitch-leveled scheduling with reduced batch size. All this is focused to implement a safe disruption-free SPF triggered by customer demand pull. It clearly shows that Lean is not a toolbox from which to select just a nice tool, Lean is a production system consisting of a tool system, or better techniques, of which every tool has to be put in place to assure a flawless production. Implementing this tool system eliminates automatically and implicitly most of Muda. However, even this model is not apt to show the interoperability of tools for a complex product manufacturing system which certainly will need to go more into detail.

The required main techniques to implement a flawless SPF of a transfer line or a manufacturing cell have been shown in Fig. 2.2. A real manufacturing environment, however, is made of several products needing several machining operations performed in different cells. These cells C_j or better shopfloor ateliers comprise usual processing-technologies such as sawing, machining, grinding, welding, heat treatment (often batch operated), surface treatment, assembly and painting. The simplest production case is the *mono*-product manufacturing, ideal for the introduction of a SPF to reduce PLT. This is done by minimizing WIP with a paced production line. To guarantee the correct takt of the line, the already mentioned techniques such as 5S, standard work, TPM, Jidoka, balancing operations have to be put in place. When *multiple* products are manufactured within the same cell (mixed-product cellular manufacturing), still maintaining a SPF, a further complication has to be mastered. Indeed, the batches B_k of a product k have to be sized to the takt rate TR_k and the workstation turnover time WTT_j of the cell C_j if a JIT delivery of several products is required. The applied techniques for this purpose are SMED and Heijunka box scheduling as well as cell design for the correct staffing. The production situation is often a complex-product manufacturing environment comprising different processing-technologies in different cells. In this further extended complexity, several manufacturing cells are linked together via strategic buffers, called supermarkets. Such buffers decouple the non-synchronized demand (D) of the downstream cell to the supply (S) of the upstream cell due to different cycle times (CT) of operations between the cells. The conveying of raw material to the cells, i.e. the internal logistics, can be implemented via optimized milk-runs as we will see. The replenishment of the supermarkets is self-controlled via Kanban, triggering the production when a stockout approaches. And finally, the requirements to be observed for a customer on-time-delivery (OTD), is that the smallest exit rate ER_j of all cells C_j has to be greater than the required customer imposed TR, and that the process lead time PLT_Z of the last, i.e. of the customer "visible" processing step Z—corresponding to the manufacturing order entry point—have to be shorter than the expected delivery time (EDT) of the customer. These are the necessary and sufficient conditions for an OTD. This means finally fulfilling a customer JIT supply. Such an extended model is shown in Fig. 2.3 as well as in [1, 8] which reflects the mathematical full induction or backward-chaining logic (i.e. from the individual to the general view), going from the mono-product manufacturing, via the multi-product manufacturing to the complex-product manufacturing. It represents the increased complexity related to in-house logistics.

All these interactions are shown in the cognitive model of Fig. 2.3, a comprehensive, at the same time schematically simplified view of the modular construction

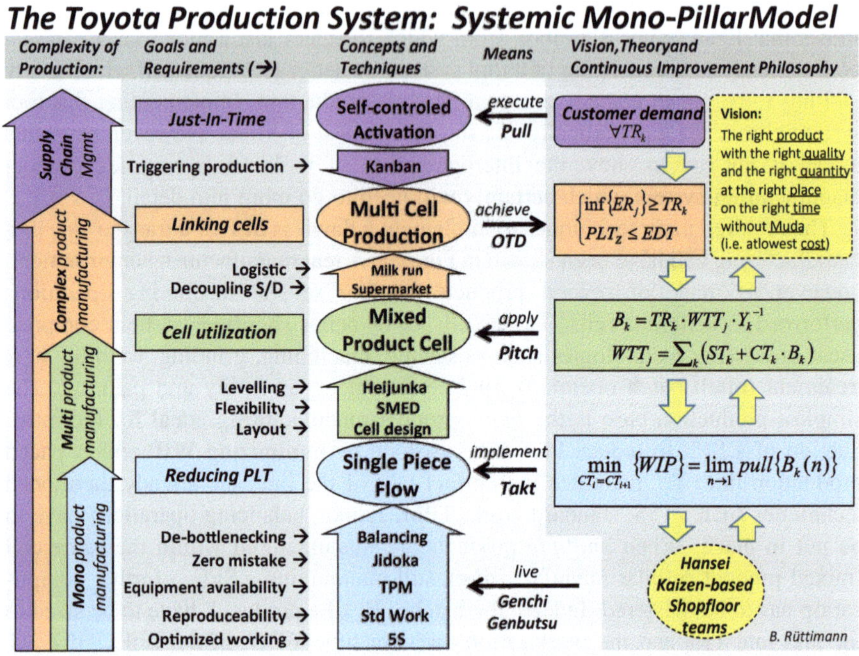

Fig. 2.3 The comprehensive mono-pillar Lean model showing the synergic mechanism of the TPS [adapted from 1, 8]

of the TPS-Lean model, showing also the rationale for each logical manufacturing complexity. It clearly states the goal of the concepts and which technique needs to be applied in order to satisfy the requirements to achieve the overall goal. In addition, the shopfloor continuous improvement is shown too (Kaizen teams), which represents the daily small improvements on all stages. Indeed, the final vision of "the right product, with the right quality, and the right quantity, at the right place, on the right time, without Muda" needs the implementation of all TPS techniques which transform the underlying theory into action, will implicitly lead to eliminate Muda. Western companies probably have the impulse to add "at the lowest cost" to this final vision, what, however, is not necessary, since achieving this vision implicitly leads to lowest cost. The new TPS model of Fig. 2.3 stands at the base of the next chapters.

Figure 2.3 exemplarily shows in a simplified manner the multiple tasks of producing within a lean-optimized, complex manufacturing environment. It shows synergic concepts and techniques and how they work together (also simplified, but explicitly modeled). In Fig. 2.3 the word tool has by purpose been replaced by technique to emphasize the aspect of necessary requirement to be used; indeed, a tool may be used or not, a technique has more relevance with regard to the "how" the theory is applied. Implementing all these concepts with the available techniques will automatically reduce the major part of waste in form of transport, inventories and WIP, waiting time which is mainly queuing time, overproduction, and quality issues. The TPS is therefore an implicit way to reduce much of Muda simply by implementing the TPS techniques and elevating manufacturing performance to the highest score. On the contrary, the Lean approach, as the reduced Western approach of VSM (VSM which is not a TPS tool, but was perfected by the Americans [9]) is an explicit way to show and to eliminate Muda in some way, this is especially the case in service companies. It is now evident, that the often applied Lean approach is not completely identically with the TPS, despite Lean and TPS are considered to be synonyms.

Indeed, TPS is an organically grown system having nearly attained perfection with Toyota whereas Lean thinking—being an emulation of the TPS—comprises the explicit transformation from B&Q to SPF as well as the explicit focus on waste reduction in order to improve the manufacturing system. Although, performing a VSM exercise, showing Muda and recursive loops "to lean it up", is only a limited view of Lean, but often applied, nevertheless, it is for sure the ideal approach for starting the Lean journey to achieve OPEX (operational excellence). Indeed, VSM is one of the most powerful tools of Lean to visualize and therefore to understand basic manufacturing principles, how the manufacturing system works and to begin the Lean transformation of Western companies not only in industries, but also in services.

The evidence is appearing that from "thinking lean" chasing Muda and by the effective communication to reduce Muda, is a target-hitting powerful marketing slogan, finally to "lean-up" everything. On the contrary, the TPS bears a "hidden" but solid and perfect production theory which contrasts the western B&Q approach going beyond explicit Muda reduction. Therefore, to explain Lean with trivial "Muda reduction" by so-called Lean consultants is indeed far too limited and should be avoided; the comprehensive sense of Lean including the systemic theory aspects should be divulged too. Further, in order not to banalize the proven TPS with saying that Lean equals Muda eradication, it is preferable to describe or better to **define Lean as a "Kaizen-based JIT production"**. This definition covers the dichotomic nature of Lean referring to the implicit Muda reduction by saying how it is implemented (by JIT production) as well as the strive for perfection by saying how it is managed (by Kaizen).

2.3 Physical Analogies to Model Production Systems

Mathematics is a divine science. Many phenomena in nature find their mirror image in mathematical equations. Moreover, many physical phenomena have a similar mathematical structure as if being mathematics itself a natural phenomenon; indeed, e.g. the discharge of a capacitor or the emptying of a level-controlled water reservoir are represented both with differential equations of the same structure. Let us take the widely used modeling of hydraulic engineering to transform the conceptual cause-effect modeling of system dynamics theory.

Figure 2.4 shows a simple cause-effect model of the filling and emptying of a water reservoir. The negative feedback loop shows the stabilizing character of the system controlling the level of the reservoir; the higher the level y, the higher the outflow, i.e. stabilizing the system and not blowing-up. This cause-effect relation of system dynamics can be translated into a hydraulic model of a reservoir with time-variant or time-invariant inflow q_{in} and variable outflow q_{out} as shown in the picture by the level-controlled valve. Indeed, $\lambda = 1/T$ is the "pace" of control regulating the effluent, where T, such as RC in an electronic circuit, represents the average time of adaptation. Based on the description of hydraulic systems, the dynamic of water level y is modeled by the following simple differential equation, which describes the temporal evolution of the level of water with adaptive emptying λy, i.e. controlled by the level itself and the not further specified inflow function $q_{in} = q(t)$ of the reservoir of system's exogenous nature.

$$\frac{dy}{dt} = q(t) - \lambda y$$

$$\dot{y} \cdot e^{\frac{t}{T}} = q(t) \cdot e^{\frac{t}{T}} - \frac{y}{T} \cdot e^{\frac{t}{T}}$$

$$\dot{y} \cdot e^{\frac{t}{T}} + \frac{y}{T} \cdot e^{\frac{t}{T}} = q(t) \cdot e^{\frac{t}{T}}$$

$$\frac{d}{dt}\left(y \cdot e^{\frac{t}{T}}\right) = q(t) \cdot e^{\frac{t}{T}}$$

$$y \cdot e^{\frac{t}{T}} - y_0 = \int_0^t q(\tau) \cdot e^{\frac{\tau}{T}} \cdot d\tau$$

$$y = y_0 \cdot e^{-\frac{t}{T}} + e^{-\frac{t}{T}} \cdot \int_0^t q(\tau) \cdot e^{\frac{\tau}{T}} \cdot d\tau$$

If we assume a constant inflow rate $q(\tau) = q_0$ then we have an asymptotic behavior of y approaching the equilibrium $y = q_0 T$ as we can see in the following calculations where T represents the average adaptation time

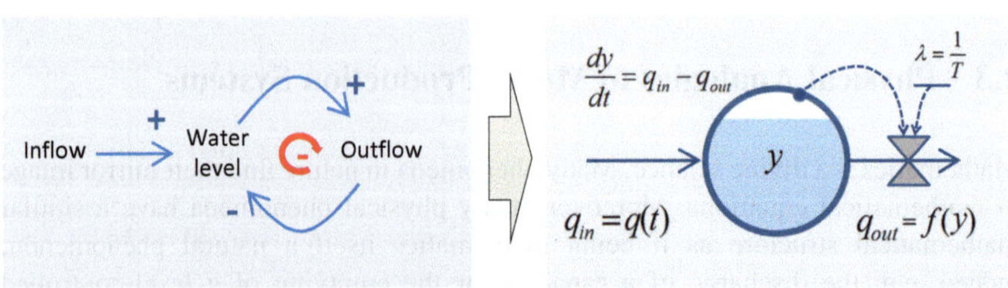

Fig. 2.4 Cause-effect diagram modeling the dynamics of a water reservoir with stabilizing, negative feedback loop translated into a hydraulic model serving later as analogy for manufacturing systems to derive paradigmatically the laws of WIP formation

$$y(t) - y_0 \cdot e^{-\frac{t}{T}} = e^{-\frac{t}{T}} \cdot q_0 \cdot \left[T \cdot e^{\frac{\tau}{T}}\right]_o^t = e^{-\frac{t}{T}} \cdot q_0 \cdot T \cdot \left(e^{\frac{t}{T}} - 1\right) = q_0 T\left(1 - e^{-\frac{t}{T}}\right)$$

$$\lim_{t \to \infty} y(t) = \lim_{t \to \infty} \left\{y_0 \cdot e^{-\frac{t}{T}} + q_0 T\left(1 - e^{-\frac{t}{T}}\right)\right\} = q_0 \cdot T$$

Figuratively, the differential equation explains the variation of the level by the net balance of inflow and outflow within an infinitesimal time interval. Although in manufacturing systems the exit rate ER is usually not controlled directly by the WIP, WIP-based ER is still the domain of management decision taken for staffing a manufacturing unit to increase ER. For our purpose, we will use a similar analogy to enounce and formalize later the dynamic of WIP formation, corresponding to the water level y of the hydraulic model. However, we will not have to deal with differential equations but for our purpose algebraic equations will be sufficient.

We can go further by entering the topic of vector spaces for modeling, becoming even more abstract. Let us define a vector space $\underline{q}$ in RxRxR, e.g. an electrical field

$$\vec{q} = \begin{pmatrix} u(x, y, z) \\ v(x, y, z) \\ w(x, y, z) \end{pmatrix}$$

and take the concept of divergence, concept which might have been forgotten if you have grey hairs. The divergence is defined for vector fields. Written with the Nabla operator notation applied to $\underline{q}$ it is

$$div\vec{q} = \nabla\vec{q} = \frac{\partial u}{\partial x} + \frac{\partial v}{\partial y} + \frac{\partial w}{\partial z}$$

which returns a scalar value regarding the balance of the flow of an infinitesimal volume Q(x,y,z) in a vector field. If $div\underline{q} = 0$ the point Q is neither a source ($div\underline{q} > 0$) nor a sink ($div\underline{q} < 0$). If a divergence-free vector field is also stationary, then, with a little bit of imagination, we could allegorically define a fast takted SPF (i.e. a continuous flow) with the notation

$$\left\{\lim_{CT_i \to 0} SPF(CT_i) : |\nabla\vec{q} = 0\right\}$$

Indeed allegorically, because a production system does not show the characteristics of a vector field. Nevertheless, to continue, $div\underline{q} < 0$ would represent a time trap, building up WIP and introducing a delay. You have not to fear, we will maintain our promise to limit math and not become too abstract. However, this example of divergence just shows intuitively the concept of flow and how certain concepts such as equal arrival and departure rate, i.e. WIP variation is equal to zero, can be explained by physical analogy.

Usually, the goals of Lean are written in the "roof" of the two-pillar TPS model such as in Fig. 2.1. When we are talking about the goals of the TPS we intend here the governing goals and not the Hoshin Kanri policy deployment cascade. The goals stated in the roof of the two-pillar TPS model are multiple. In mathematics, this equals to optimize a multi-objective function. In operations research and economics it leads to the concept of Pareto optimality. Pareto optimality applies in the context of concurrent multi-objective maximization (or minimization) of functions. A solution is called Pareto optimal, or Pareto efficient, when any increase of one objective function is made to the detriment of another. The optimum is achieved when this condition is attained. Although the goals and mindset of Lean have already largely been divulged and are commonly known, I prefer to add a further model of the Lean objective system leading to Pareto stability calling it the Lean Governing Principles. Indeed, we could apply a further paradigmatic physical analogy, the one of the thermodynamic postulates, to explain the regal-like Lean reigning goals; they show that also a production system is subjected to follow a logic of "divine" rational (Fig. 2.5). This framework of Lean Governing Principles shown in Fig. 2.5 is topologically closed, i.e. it is comprehensive according to the definition of mathematical topology. Indeed, the properties of compactness and connectivity are given in an extended interpretation. Such as the thermodynamic postulates govern the evolution of thermodynamic processes, in Lean we talk about

- leveling demand to obtain a steady-state dynamic equilibrium represented by a flow;
- optimizing resource allocation which is equivalent to reduce cost or increase productivity;
- speeding-up processes which is equivalent to reduce WIP; and finally
- striving to attain zero defects, which has to be the governing rationale of each production department and plant.

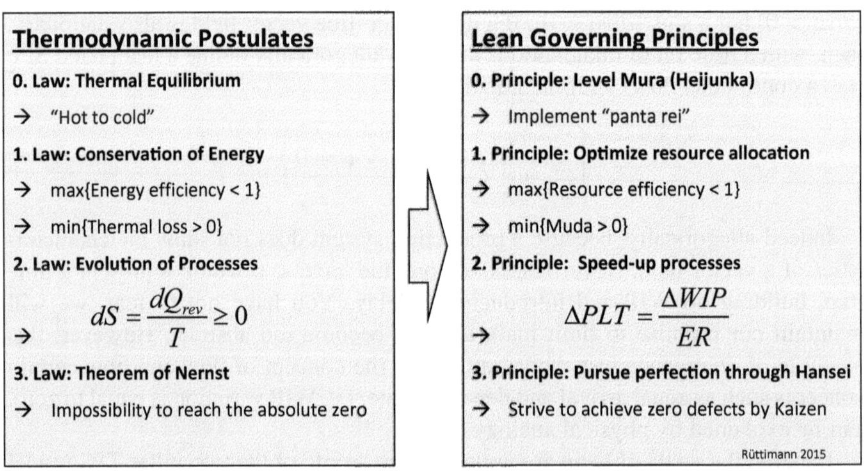

Fig. 2.5 Comparative analogy of the Lean Governing Principles to the thermodynamic laws [adapted from 6]

The Toyota Way puts a lot of emphasis on the cultural aspects how a company is managed, how it empowers people, how decisions are taken and implemented. These Toyota credo have been enunciated in 14 principles by Liker [3]. According to the more tangible-driven topics of this book, we will focus on the technical manufacturing aspects. In the following, we will consider our production system as being a physical system according to the Lean Governing Principles framework of Fig. 2.5, possible to be modeled mathematically by an appropriate analogy, discussing the systemic operability presented in the new Lean model of Fig. 2.3. Therefore, in this book we will deal especially with the principles 0 and 2 to confer them an additional attention. The first and third principles have been already widely discussed in other books e.g. [3], but also the 0 and 2nd one, although they are not specifically called governing principles. I want to clarify, despite the semblance of a Pareto-similar optimality criteria of the Lean Governing Principles might be given, it is noticeable to highlight that they bear a tautological nature, the principles not being in contrast with each another but leading to the confluence of a common synergic objective of perfection. Please also note, the analogies between thermodynamics and Lean are surprising but they have been constructed artificially to match figuratively, especially the phenotypic semblance of Clausius's entropy equation by adapting Little's original law to discrete incremental variation. Indeed, whereas the entropy equation of the second thermodynamic postulate is based on the property of the extensive heat variable Q linked to its intensive variable temperature T, the variables WIP and ER in the second Lean principle, explaining PLT of a production system, present an indirect connection and their intrinsic "physics" properties are independent.

Just for clarification, the term production system is the most generalized concept of modeling input-output relations in an economic system of resource transformation. Whereas on the one hand a manufacturing system has more the connotation of a shopfloor operational system of physical transformation, best represented by an assembly operation, on the other hand, a processing system is best represented by a chemical process. The distinction between manufacturing operations and processing operations may become fuzzy in certain production systems, also because of mixed systems (production of a tissue substrate becoming paper at a first step and cutting of paper rolls into paper sheets in a second step). In the following, for simplicity reasons, we will deal primarily with operations of manufacturing type.

References and Selected Readings

1. Rüttimann, B., Stöckli, M.: Going beyond triviality: The Toyota production system—lean manufacturing beyond Muda and Kaizen. J. Serv. Sci. Manag. **9**, 140–149 (2016)
2. Ohno, T.: Toyota Production System—Beyond Large Scale Production. Productivity Press, New York (1988)
3. Liker, J.: The Toyota Way, 14 Management Principles from the World's Greatest Manufacturer. McGraw-Hill, New York (2004)

4. Womack, J.P., Jones, D.T.: Lean Thinking. Free Press, New York (2003)
5. Rüttimann, B., Walder, H., Adam, M., Stöckli, M.: Lean Six Sigma in der Schweiz—Explorative Studie zur Standortbestimmung, inspire/ETH, SISE (2012)
6. Rüttimann, B., Wegener, K.: Einführung in die Methoden von Lean Manufacturing und Six Sigma Quality Management, ETH Tools-IV Kurs, Lecturing notes HS2014, D-MAVT (2014)
7. Rüttimann, B.G.: Discourse about linear programming and lean manufacturing: Two different approaches with a similar, converging rational. J. Serv. Sci. Manag. **8**, 85–91 (2015)
8. Rüttimann, B.: Von Lean zu Industrie 4.0 – eine Evolution? Von einer visionären Idee zum realen Verständnis, Presentation held at: Fertigungstechnisches Kolloquium, ETH Zürich, November 26, 2015 (2015)
9. Rother, M., Shook, J.: Learning to See. LEI/Cambridge Center, Cambridge (2003)

Chapter 3
Preliminary Concepts, Definitions, and Basic Production Laws

In the following sections we develop some taxonomies regarding production, manufacturing and transfer principles. We introduce the queuing theory and related WIP formation defining the bottleneck and the general production requirements for OTD, constituting the cardinal concepts of production.

3.1 Components of a Production System

A production system can be best represented by a value stream map (VSM) as it has been described by [1] to which we refer. The VSM is a process representation of medium aggregation level as shown in Fig. 3.1; other current process representations are on the one hand high level schemes of input-process-output, i.e. IPO-type (low resolution), or on the other hand, cell design of low level with detailed mapping of activities, i.e. with high resolution of detail representation. VSM shows the following main components:

- demand and its characteristic requirements (takt rate TR and expected delivery time EDT)
- sequence of manufacturing process (data quantified process steps and material flow)
- information about supply modality (e.g. frequency of inventory replenishment)
- information about production planning (dispatching of manufacturing orders).

The VSM synthetically shows process steps, material flow, information flow, and process metrics. It allows to representing the actual process, how it is governed, how it is performing, discovering hidden improvement potentials, i.e. the visualization of obvious Muda. In Fig. 3.1 we highlight some of the key topics we will deal with during the next chapters such as

© Springer International Publishing AG 2018

B.G. Rüttimann, *Lean Compendium*, DOI 10.1007/978-3-319-58601-4_3

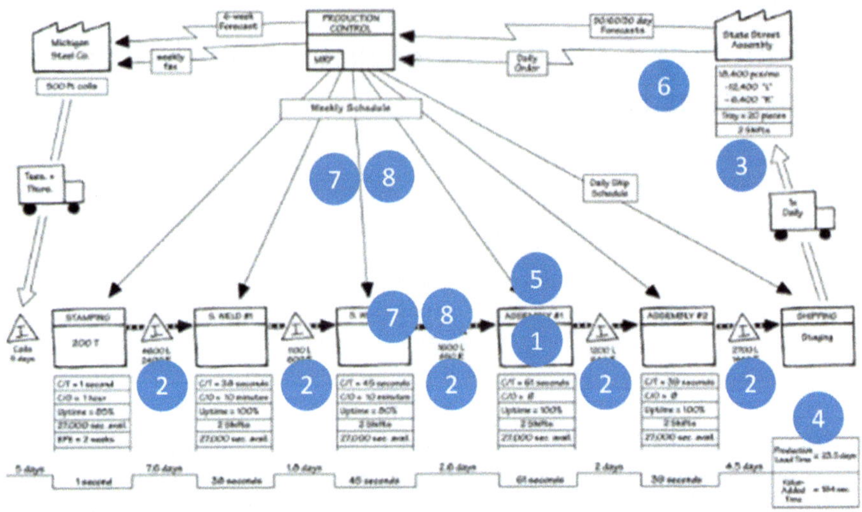

Fig. 3.1 Typical VSM [from 1] with main topics of manufacturing highlighted being treated afterwards

1. Bottleneck and its impact on the throughput (Chap. 3)
2. WIP and the implication on process lead time (Chap. 3)
3. OTD and which conditions have to be satisfied (Chap. 3)
4. PLT performance of different manufacturing modi (Chap. 4)
5. CTT (cell turnover time) reflecting the concept of bottleneck for mixed-product manufacturing cells (Chap. 5)
6. TR as the imposed manufacturing rhythm (Chap. 6)
7. Cell decoupling supermarkets (Chap. 6)
8. Kanban to link manufacturing cells (not shown explicitly on the VSM) (Chap. 7)

These topics are not only relevant to understand the TPS, but they represent also the basic concepts of every production system. We will therefore discuss them generally and then drive the attention to the aspects of the TPS.

3.2 Taxonomy of Production Principles

Still existing some confusion about current principles, such as B&Q, SPF, pull, or push, let us therefore introduce first some definitions in order to clarify and systemize the different interpretations of these principles. We will use throughout the book the definitions given in this book which may be different to other books, if ever addressed.

The production batch B_k for the product k corresponds usually to the ordered quantity requested by the customer. Technological constraints, such as e.g. a combined melting/casting equipment of 35 tons does not allow to pour a 10 ton batch, forcing to fill the melting/casting unit up to 35 tons (observe technical constraint). On the other hand, a big customer order of 200 tons may not be cast in a consecutive sequence of six identical pourings due to concomitant delivery requirements of other orders (in traditional manufacturing management environment, however, most probably it would consist of one batch), the big order of 200 tons needs to be split, despite setup costs ("washing"), and the other orders interposed in between the big order (to match delivery requirement). In such circumstances the smallest planning unit is therefore the batch with 35 tons, which usually is called pitch and the backlog is leveled in a Heijunka box (see Sect. 5.1). Different than in processing-type of operations, in manufacturing-type of operation realities, things are simpler and allow more flexibility, usually only restricted by the setup cost of order change. Therefore, a production batch of B_k pieces in a manufacturing type of operations could be fabricated indifferently in a standard or customized batch size. However, how the batch is fabricated may vary considerably. Indeed, the pieces of the batch $B_k(n)$ can be pushed downstream through the several fabrication steps either by

$$B_k(1) = B_k \cdot 1$$

i.e. "make-one, move-one", which corresponds to the intuitive concept of SPF, also called one-piece-flow or sometimes simply flow, or

$$B_k(B_k) = 1 \cdot B_k$$

i.e. "make-all, move-all", which corresponds to the concept of B&Q; note, "make all" means here one after one. It might be possible, due to logistics restrictions, that pieces are linked to the pallet size, and that the grouped transfer unit n is not unitary (please note, sub-batches of size n moved as a whole but each piece is manufactured in n consecutive cycles)

$$B_k(n) = l \cdot n \tag{3.1}$$

i.e. "make-n, move-n", which corresponds to an intermediate modus with some flow characteristics; we could call it n-piece-flow (nPF). The more n approaches one and where the multiplier l approaches the lot size, the more it corresponds to a SPF characteristics; e.g. for $n = 4$, this "quartet" is handled SPF-like, i.e. "make-four, move-four". This intermediate nPF modus may become operational when a logistic transportation unit of transferring the pieces from one operation to the next has for handling reasons a capacity greater than one. Please, do not confound with the case a die tool has multiple cavities c; indeed, in this case c pieces at each cycle time are made which is a capacity topic (see later); we would name that "make-n(c), move-n". We will call the way how the batch runs through the different operations, i.e. SPF, nPF, or B&Q, the **transfer principles**.

On the other hand, we can distinguish the ways how the transfer is triggered, i.e. which is the rule applied to signal to move the pieces from one station to the other station more downstream; we will call these the **manufacturing principles**. The two main manufacturing principles are:

- "push", i.e. the pieces are literally pushed from one operation step (upstream) to the next operation (downstream) as soon as they are ready to me moved (either SPF or B&Q wise), or
- "pull", i.e. the pieces are called-off from the upstream operation by the downstream operation; as we will see, also mixed principles "push-pull" may co-exist within the same value stream.

Further, the two main **production principles** are:

- "make-to-stock", i.e. production is made to replenish a logistics or consignment inventory, which suits usually off-shelf standard products, or
- "make-to-order", i.e. production is made on demand originated by a specific customer order, which may be for customized but it suits also for standard low runners.

Despite the two production principles seem to be quite different, the manufacturing logic is basically the same and the reasoning applied in the following is valid for both production principles.

An additional characterization of the manufacturing policy are the **scheduling principles**, which can be divided into the following kind of natural or campaign scheduling:

- first in first out (FIFO), sometimes also called first come first served (FCFS)
- last in first out (LIFO),
- earliest due date (EDD)
- shortest process time (SPT)
- general FIFO with preferential orders to observe OTD (priority FIFO)

or the Lean-type TPS approach

- Heijunka box pitch-leveled according to random-based consumption-triggered scheduling of a deterministic mix.

In the following, we will assume FIFO if not specified differently.

The transfer and the manufacturing principles can be combined crosswise resulting in the contingency matrix of Fig. 3.2. The resulting four **main manufacturing modi operandi** are:

- "single piece push", on the one side we have for a single transferred unit and in takted and CT-balanced line a so-called single piece flow which we usually abbreviate SPF (also called one piece flow), best represented in the ideal case by high performance transfer lines TFL (paced assembly lines with deterministic CT imposed by the conveyor belt). But not exclusively; such a line can also be operated manually where the operator may introduce variation in assembly, leading to stochastically balanced lines with formation of WIP; these lines are sometimes also called "flow on pull", or FIFO lane in the case that small buffers are between the operations;

- "single piece pull", the non-necessarily takted, i.e. non-takted FIFO lane with the upstream operation working only when receiving a Kanban; the single piece pull (SPP) could also be called single piece handling (SPH) instead of SPF. Such types of "flow lines" with independent workstations have a bottleneck. These two modi usually are named "flow shop" organizations.
- "batch push", on the other side we have the classic batch-push, usually known as batch and queue (B&Q), generating WIP between operations; the batch usually corresponds to the order quantity. This is the most frequent manufacturing modus applied in Western production plants;
- "batch on pull", batch operated shopfloor but starting only when a Kanban occurs, therefore also called generic pull, intending to limit WIP (controlled push); these last two modi usually named "job shop" organizations.

Please note, the batch-pull is sometimes also called generic pull or Conwip (constant WIP). Indeed, viewing a process as a black box, i.e. not considering the internal manufacturing organization of the single work stations, as soon as a batch leaves the process, a new batch enters the process, being the exit of a finished batch a sort of Kanban. This input-equals-output (exit) "modus operandi" allows to stabilize the WIP and therefore the PLT as we will see (Sect. 7.1). If we consider the black box as white box, looking inside the process chain, we see that the products are pushed through the work stations; this is the reason, why generic pull systems may also be called controlled push systems. We have to stress, that the classic batch-push and the single piece pull are "pure" modi, whereas the batch on pull, just described, and the paced SPF are mixed modi. Indeed, the paced SPF is a clear single piece push, but the triggering is usually a downstream pull signal, which is the aim of a JIT Lean manufacturing line; this modus is then called "flow on pull".

In addition, notice that the taxonomy matrix of Fig. 3.2 is a rough classification; indeed, we assimilated the single piece pull to a FIFO lane and this might be controversial. Indeed, in a pull principle the FIFO organization is not natural from a genotypic view but it might be acceptable from a pure phenotypic view. We encounter a FIFO lane also in non-perfectly balanced push SPF lines, which then result in FIFO lanes with buffer capacity. But such a FIFO buffer allows to avoid to stop the whole takted line. Of course such an organization is testimony of a not yet optimized cell or transfer line design or a deliberately accepted situation to bridge best possible occurrences of imperfections. All these examples show from where the confusion of the naming might originate.

A simplified VSM-type representation of the two main manufacturing modi for the "make-to-stock" production principle is shown in Fig. 3.3. Usually the VSM shown here on the left side of Fig. 3.3 is a pre-transformation status based on B&Q and the VSM on the right side is the ideal lean solution of a post-transformation SPF carried out by a Lean transformation based on the drum-buffer-rope (DBR)

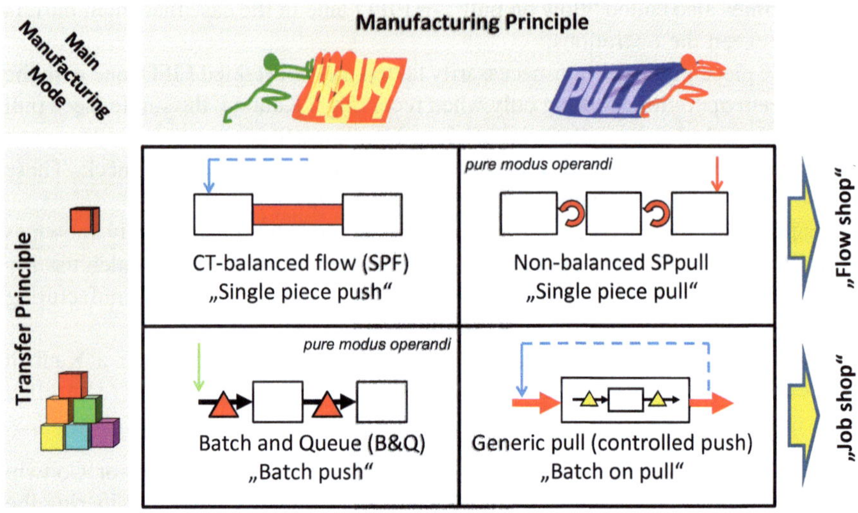

Fig. 3.2 Contingency matrix of the transfer and manufacturing principles results in four main manufacturing modi operandi

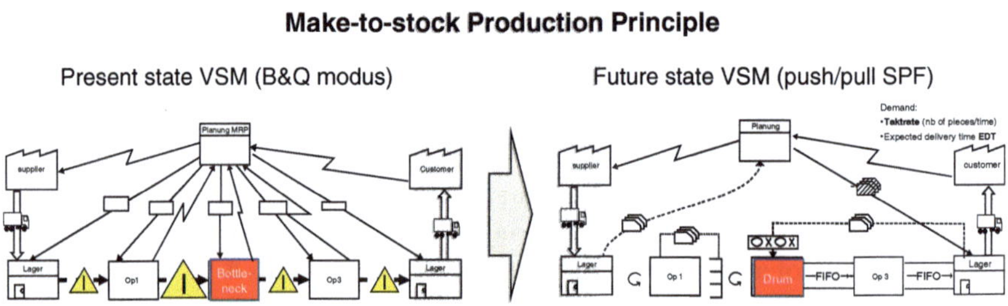

Fig. 3.3 Make-to-stock production comparison of a classic B&Q manufacturing modus with a lean-type mixed push/pull SPF [from 2]

technique which we will see later (Sect. 4.4). The "make-to-order" production principle for both manufacturing principles the B&Q push as well as the Lean push/pull principle is shown in Fig. 3.4. In the following we will deal with the behavior and performance of these two main systems.

3.3 Queuing Theory and WIP Formation

Queuing theory is a specific topic of mathematics useful to describe the dynamics and performance of workstations. A queuing system combines the aspect of an arrival process and an execution process, and the configuration of execution. It is usually described with the following notation:

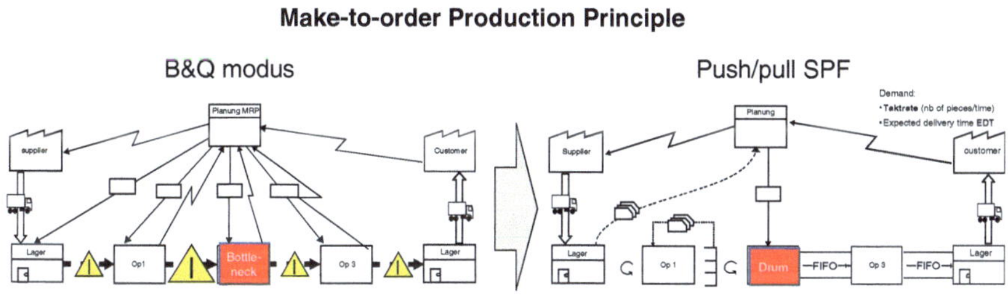

Fig. 3.4 Make-to-order production comparison for push and pull manufacturing principles

$$(X/Y/s/b)$$

where X describes the process of inter-arrival time, Y describes the process of execution time, s is the number of parallel servers, and b represents the buffer capacity. X and Y can have a D (Deterministic), an M (Markovian, i.e. exponential), or a G (General, e.g. normal, uniform, or other type) distribution. The most common systems are Markov-process derived queuing systems, i.e. based on an exponential probability density function (PDF) of type

$$f(t) = \lambda \cdot e^{-\lambda t}$$

where $t \geq 0$ and $\lambda > 0$ is the average rate, with an easy computable integral

$$F(t) = \int_0^t \lambda \cdot e^{-\lambda\tau} \cdot d\tau = \lambda \cdot \left.\frac{e^{-\lambda\tau}}{-\lambda}\right|_0^t = -e^{-\lambda\tau}\Big|_0^t$$

The integration of the PDF results into the cumulative probability distribution function

$$F(t) = P(T \leq t) = 1 - e^{-\lambda t}$$

Notice, an exponential distribution is the dualistic view, or better the inverse concept of a discrete Poisson distribution having for the probability mass function and its cumulative function respectively the expressions

$$p(k, \lambda) = \frac{\lambda^k \cdot e^{-\lambda}}{k!}$$

$$P(K, \lambda) = e^{-\lambda} \sum_{k=0}^{K} \frac{\lambda^k}{k!}$$

where $\lambda > 0$ is the average rate of occurrences and p is the probability of $k \geq 0$ number of occurrences. Whereas the exponential distribution deals with the interval time t between occurrences, i.e. continuous values, the Poisson distribution deals with the k number of occurrences in a specific time frame, i.e. integer values. Different queuing systems are imaginable, e.g.:

$$\begin{cases} a) := (G/G/s/b)\_or\_(M/G/s/b) \\ b) := (M/M/s/\infty) \\ c) := (D/D/1/\infty) \end{cases}$$

Usually, real processes are of type a); queuing theory is often comprehensively explained for processes of type b), and in the following, for simplicity of introductory didactic reasons, we will deal primarily with type c) processes at the beginning and make some excurses to the other distributions. The deterministic arrival process can be justified for paced call offs, i.e. with an imposed takt rate TR by the customers. We define the limit of a normal distribution with the standard deviation tending to zero

$$\delta = \lim_{\sigma \to 0} N\left(\mu, \sigma^2\right)$$

approaching conceptually Dirac's δ-function, as a deterministic execution process. This can be applied to ease first comprehensibility assuming further simplifications, exactly such as no variation in cycle times CT, i.e. no slowdowns, where the CT is the process time of execution to manufacture one piece; no setup time; idealized utilization rate with no aspects of server saturation, no breakdowns, etc.

For further explanations of queuing theory, and those readers more familiar with mathematics, we refer to academic text books, such as e.g. [3, 4]. Indeed, in the following we will use a more simplistic language, better comprehensible and suitable for proven production managers, as the own lecturing and consulting experience has shown. It has also to be pinpointed that queuing theory may have a higher importance in Western companies, if ever, due to the B&Q transfer principle. Indeed, the Japanese view of manufacturing is not to optimize queuing and to calculate residual queuing time, i.e. to manage queues, but to avoid long queues installing a SPF and by implementing a Heijunka box leveling to eliminate the negative effects of backlog waiting time (BWT) due to queuing of orders and batches. It is curious, but the manufacturing culture between Western companies and Toyota could not be more different having an antithetic rationale and consequent different behavior.

To avoid ambiguity and to facilitate comprehension, let us define upfront the following rates in queuing systems of various aggregation levels shown partially in Fig. 3.5 and Fig. 3.6:

- OR is the order rate at which the customer orders
- IR is the input rate at which the scheduling department releases orders to shopfloor

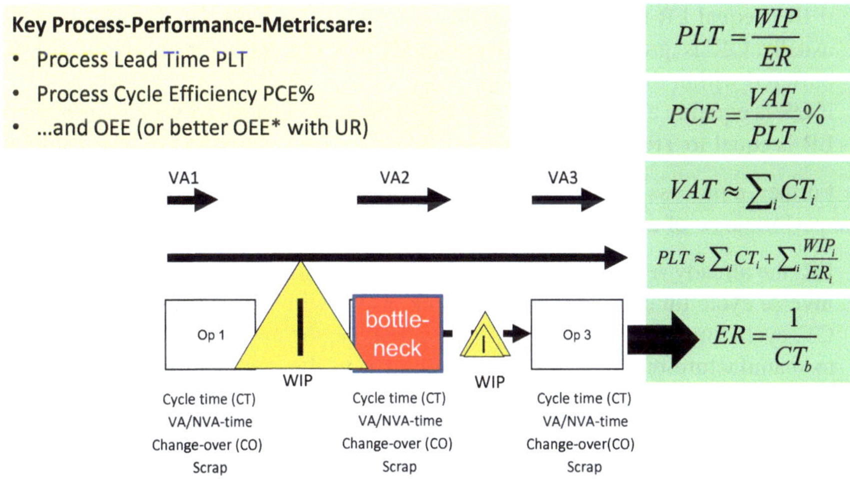

Fig. 3.5 Key process metrics characterizing the performance of a process [from 2]

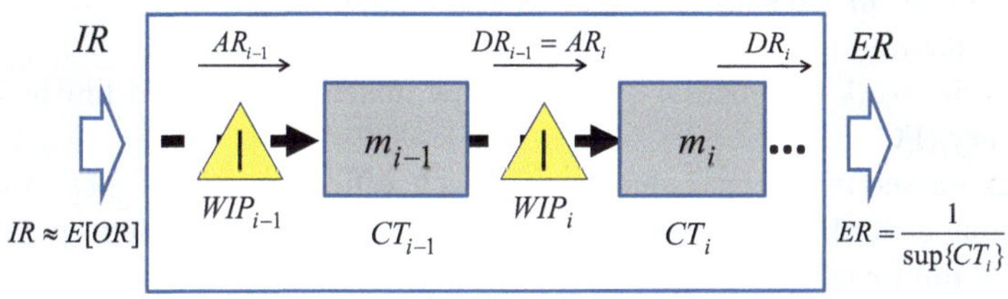

Fig. 3.6 Value stream excerpt of a manufacturing system serving to explain the Law of WIP formation where $DR_{i-1} = 1/CT_{i-1}$ and $AR_i = DR_{i-1}$

– ER is the exit rate at which the manufactured pieces leave the production process and are stocked or directly shipped to the customer. For service operations it is often called completion rate.
– AR_i is the arrival rate and DR_i is the departure rate to and from a workstation; these are the rates of pieces (or also entire batches, when specified) which arrive to and leave the machine m_i or generally an operation; it has to be said that DR_{i-1} equals AR_i for obvious reasons (see analogy $\mathrm{div}\underline{q} = 0$). For a white box approach, AR and DR are essential, for a black box approach E[OR] and ER are the relevant rates. IR has forcedly to be equal ER to have a stable process, as we will see. The OR and IR are equal to TR in customer paced orders. Please note, the breakdown of a yearly demand to a daily requirement is only possible if no bulk arrival is supposed, i.e. an even distribution of the orders is at the base. In the following, we will use OR for general arrivals of orders and IR for generally unbalanced manufacturing lines and TR rather specifically for SPF balanced lines. TR is the takt rate for rather even distributed demand of customer orders. We anticipate some basic concepts:

- if IR is equal ER the process is stable
- usually OR is greater or smaller than IR, i.e. the high variability of OR is buffered within the order backlog
- if OR is equal to an explicit TR then try to fabricate in a paced way assuring that ER is equal to TR.

Further, a process is characterized by the following possible metrics to measure the performance of a process shown in Fig. 3.5:

- ER, the already seen exit rate being the specific capacity of the line given by the inverse cycle time CT_b of the bottleneck, as we will see
- CT, cycle time is the time a single operation, i.e. the machine or operator, needs to manufacture one piece before handling it to the next station, generally not including setup time
- PLT is the process lead time a piece needs to go through the process including queuing time. Please note, sometimes this is also called process cycle time or simply cycle time causing confusion
- VA or VAT is the value add time for which the customer is willing to pay and corresponds generally to the "touch" time; we will assume for simplicity that the touch time is all VAT
- NVA is the non-value add time, as waiting time, setup, rework
- WIP is the work in process, excluding raw material (RM) and finished goods inventory (FGI)
- PCE expressed in percent is the process cycle efficiency, sometimes called lean indicator, given by the ratio of VAT and PLT, showing the latent potential of residual improvement.

Let us now suppose to have a manufacturing process consisting of m production steps, each step, or operation with its own cycle time CT_i and, as already stated, ideally with zero setup time. In addition, let us suppose that each operation has only one server, i.e. machine or operator when not specified differently, and at each machine or operation cycle only one piece is handled (die with one cavity). Let us further suppose to fabricate different products k on this manufacturing line or cell with not explicitly determined production batch for the product k (e.g. annual quantity). As a consequence, the CT_i should be written more precisely as $CT_{m|k}$ which associates to each different product k its corresponding cycle time at each machine m (or operation) of the manufacturing line. A manufacturing cell can therefore be represented by the matrix **A** = $[CT_{mk}]$ shown in Eq. (3.2) of non-necessary quadratic order m*k in a linear optimization exercise. Matrix representations of manufacturing cells, such as shown in Eq. (3.2), are very useful also in real manufacturing world, and hence reported here.

$$A = \begin{vmatrix} CT_{11} & CT_{12} & \dots & CT_{1k} \\ CT_{21} & CT_{22} & \dots & CT_{2k} \\ \dots & \dots & \dots & \dots \\ CT_{m1} & CT_{m2} & \dots & CT_{mk} \end{vmatrix} \tag{3.2}$$

A linear optimization exercise can be described by an objective function to be optimized and a set of restrictions. The restrictions of type $\mathbf{Ax} \leq \mathbf{b}$ on the one hand, where a single element CT_{mk} of $\mathbf{A}$ denotes the unitary specific absorbtion of the machine m by the product k, and on the other hand the non-negativity assumption of $\mathbf{x}$. The vector $\mathbf{b}$ denotes the absolute capacity, and $\mathbf{x}$ is the vector of the product mix with $\mathbf{x^*}$ denoting the solution vector of the optimal mix to maximize the objective function $z=c^T x$ [5]. In the case of the vector $\mathbf{c}$ denoting specific margin contributions, Linear Programming (LP) systems of type

$$\begin{cases} \max\{z\} = \max\{c^T x\} \\ A \cdot x \leq b \\ x \geq 0 \end{cases} \tag{3.3}$$

allows to maximize a given production system but are not especially apt to model Lean type of manufacturing systems [5]. In LP type exercises represented by Eq. (3.3) the vector $\mathbf{b}$ is said to be a bottleneck when the sign of equality applies in the restriction $\mathbf{Ax} \leq \mathbf{b}$ whereas in LM the element $\sup[CT_{mk}]$ of matrix $\mathbf{A}$ is the bottleneck [5]. The optimal solution of equation system (3.3) is

$$z_{opt} = \left\{c^T x^* \middle| x^* \in x\right\}$$

Please note, we will not use the mathematical operators *inf, sup, min, max* according to the diction of set theory, but using *min* and *max* for indicating optimization operators in the sense of applying first order condition whereas *inf* and *sup* return the lowest and highest value of a finite set of elements; this is slightly different than the strict mathematical interpretation but suits better to our need. Linear Programming (by the way, the diction Linear Programming instead of Linear Optimization originated from the fact that originally it has been applied to solve scheduling problems) is now losing its applicability in takted customer systems, i.e. with customer imposed mix and Heijunka-pitch scheduled sequence.

Let us further define the key performance drivers of a process. We will define the performance of a process limited to three indicators, i.e. to the exit rate ER, the work in process WIP, and finally the ultimate key performance indicator of a process, the process lead time PLT. The ER expressed in pieces per time, which corresponds to the commonly known completion rate λ ("Poissonian view") of a service provider, represents the specific capacity of the process and results from the bottleneck operation with CT_b ("Markovian view") reflected by the longest CT_i of the manufacturing process shown in Eq. (3.4a). Remember, the concept of bottleneck is defined differently in LP and in LM [5]. The absolute capacity depends from the number of work shifts. That means, that the specific capacity of a process, the exit rate ER, is given by the ER_b of the bottleneck (Eq. 3.4b), which formal theory has been discussed e.g. in [5]. Please note, according to our definition, the ER_b corresponds to the DR of the bottleneck, being the bottleneck somewhere within the process, but the importance of the DR of the bottleneck determining the DR of the process, i.e. the ER, we will call it also ER.

$$\begin{cases} CT_b = \sup\{CT_i\} \\ ER = ER_b = \dfrac{1}{CT_b} \end{cases} \qquad (3.4a, 3.4b)$$

Equation (3.4b) links the "Poissonian view" of ER to the "Markovian view" of CT. Equation system (3.4) represents the **definition of bottleneck**, saying that the exit rate of the process is given by the cycle time of the bottleneck. The slowest process step is called the bottleneck of a process. This definition implies, that every process has a bottleneck and that it exists only one bottleneck in a process. Equation system (3.4) allows to enunciate now the

Theorem of Throughput (or Bottleneck Theorem)

Given is a sequence of production steps with each process step having a deterministic but different cycle time CT. The maximum throughput, i.e. the maximum exit rate ER, of a process is given by the slowest process step, i.e. the process step with the longest cycle time CT; this process step is called bottleneck.

First Corollary to the Theorem of Throughput (Corollary of Bottleneck Uniqueness)

Based on the definition of bottleneck, every process has one, and only one bottleneck.

Second Corollary to the Theorem of Throughput (Corollary of Bottleneck Time-invariance)

For a mono-product cell or transfer line, if no structural changes are made to the cell or line, the bottleneck is a time-invariant property of the process. Based on the intrinsic definition of bottleneck, only by applying changes to the present bottleneck by reducing CT or by mix change may generate a different, new bottleneck.

The Theorem of Throughput shows, that the usual Western focus on the most expensive equipment is wrong; the main attention should be put on the throughput-limiting equipment, i.e. the bottleneck. But it has to be added, it would be irrational to let an inexpensive machine to become the bottleneck of the line.

This theorem is valid for a process manufacturing a single product; for multi-product lines and cells, the definition of bottleneck has to be slightly adapted (see Chap. 5). The PLT, expressed in time units, represents the timespan to manufacture one piece according to the applied transfer principle, i.e. the time one piece of the batch needs to transit the line. We will introduce a new, additional metric, precisely the manufacturing lead time (MLT) expressed also in time units, but in this case it represents the timespan to manufacture the whole batch B_k according to the applied transfer principle, i.e. the necessary time when the first piece enters the process and the last piece of the batch leaves the process. This is of special interest in B&Q operations with long changeover times, where the order is manufactured and delivered in one batch. To develop that, we will neither base on the aggregated deductive approach of **Little's Law** e.g. [3], directly derived from queuing theory

$$PLT = \frac{WIP(t)}{ER} \tag{3.5}$$

where PLT is the time a single unit of the batch needs to transit the process, and WIP(t) denotes the work in process at the instant t,—nor will we base on the apparent presumed correct inductive approach logic of explicit addition of processing time CT and waiting time given by the visible WIP_i in front of the operation m_i with CT_i (i.e. ER_i)

$$PLT \approx \sum_i CT_i + \sum_i \frac{WIP_i}{ER_i} \tag{3.6}$$

We will apply a more general, comprehensive approach to develop MLT. The PLT returns in both cases of Eqs. (3.5) and (3.6) the approximate, yes it is only approximate as we will see, necessary time that one piece, entering the line, needs to transit the whole manufacturing process. Please note, the WIP(t) of Eq. (3.5), but also WIP_i of Eq. (3.6), represent the values of a snapshot and may vary in time according to the load; see below the new enunciated Law of WIP of equation system (3.7). We anticipate, PLT will remain stable in a steady-state regime, it will vary during transient and in a constraint system, i.e. a system of not steady state. The necessary time that the batch B_k needs to transit the manufacturing line, i.e. MLT, of course depends also on the batch size itself. To have an unbiased comparison, we will consider first a line with only one product, i.e. no other product than "k" is in the manufacturing process, i.e. we will deal with the column view of the matrix **A** which generally has different $CT_{m|k}$ or, in the optimized case of a balanced cell, equal $CT_{m|k}$ for all m in the column k, corresponding to a mono product manufacturing cell [2, 5]. Secondly, we will expand this special case to the general case of the multi-product manufacturing cell, i.e. the whole view of the matrix **A** (see Chap. 5).

The WIP is the cardinal variable determining PLT as Eq. (3.5) reveals. We can give the following **definition of WIP**. As soon as a customer order is transformed in a manufacturing order and has entered the manufacturing process it becomes WIP. It remains WIP until it has left the last operation and is put to FGI (finished goods inventory) for shipping. This definition leads to Eq. (3.7a). Therefore, the FGI does not belong to WIP according to that definition and neither the RMI (raw material inventory). The amount of WIP is deterministically influenced by the manufacturing policy (e.g. scheduling principle, batch size and applied principle of SPF or B&Q) as well as structurally influenced by the capacity of the bottleneck and randomly influenced by variation of CT or arrival times. The Theory of Constraints (TOC) and the importance of the bottleneck have been described allegorically by Goldratt [6, 7] but it was not formalized figuratively at that time, and it has never been completely so far. The concept of bottleneck in LM has been comparatively analyzed to LP problems and its concept of "shadow price" in [5]. However, LP does not consider queuing, at least in the simplest case.

The inventory characterized by WIP_i shown in Fig. 3.6 and the machine characterized by CT_i represents a queuing unit i represented by the queuing system (X/Y/1). Let us enounce and formalize the dynamics of WIP formation, corresponding to the analogy of the water level y of the hydraulic model of Fig. 2.4. Figure 3.6 also shows the analogy of the hydraulic model for a manufacturing system, which WIP dynamic can be described by the exhaustive equation system (3.7a–3.7c) claiming it as the **Law of WIP** formation

$$\begin{cases} WIP = \sum_i WIP_i \\ \dfrac{\partial WIP_i}{\partial t} > 0 : CT_i > CT_{i-1} \\ \dfrac{\partial WIP}{\partial t} = E[IR] - E[ER] \approx \dfrac{1}{TT} - \dfrac{1}{CT_b} \end{cases} \quad (3.7a - 3.7c)$$

where E[IR] represents the average of the input rate, i.e. the manufacturing orders released to shopfloor which should also average in the medium run the customer order rate OR, i.e. the incoming arrival rate AR of orders, or the demand takt rate TR imposed by the customer in a repetitive business. The TR can indifferently be expressed also by its dual time dimension which corresponds to its takt time TT, i.e. the inverse of the TR. The non-negativity requirement of CT is not stated explicitly in equation system (3.7) but assumed implicitly. The TR is a concept of LM, i.e. a quasi-deterministic idealized concept of more realistic Markovian-distributed arrivals. E[ER] is the average exit rate of the process, which usually is quite stable given by the CT_b of the bottleneck. The partial derivative in equation system (3.7) is explainable by the fact that the WIP-change is not depending explicitly by time but implicitly also by other variables (e.g. variation of CT due to different products or failures). Equation (3.7c) remembers the balance of a hydraulic water reservoir seen at the beginning. For a Lean takted SPF line, being E[OR] = TR and showing ER ideally no variability, Eq. (3.7c) becomes

$$\frac{dWIP}{dt} = TR - ER = \frac{1}{TT} - \frac{1}{CT_b} = \frac{\frac{CT_b}{TT} - 1}{CT_b} \approx \ln \frac{CT_b}{TT} \quad (3.8)$$

If CT_b shows a limited variability of +/−20% of the TT, the variation of WIP can be approximated by the natural logarithm of the fraction of the two times. If Eq. (3.8) is set to zero, i.e. no change of WIP is observed, this requires $CT_b = TT$ or in the dual writing mode ER = TR, we get Eq. (3.9) which reflects the fulfillment of the weak WIP stationary requirement and is one of the requirements for a JIT-based manufacturing system, as we will see in Sect. 6.4.

$$\frac{dWIP}{dt} = TR - ER = 0 \tag{3.9}$$

If in Eq. (3.9) the equal sign is substituted by the greater than zero sign, the process is not stationary, "blowing-up" and never be able to reach the steady-state condition. In real business, investment in capacity will be the consequence. The equation system (3.7) models the formation of WIP within a production line and is at the base of the further discussion for unbalanced lines as well as lines with capacity constraints. We **define here a constraint** as a manufacturing step not being able to process the E[IR] (see below). WIP is a result deriving on the one hand by the applied transfer principle (SPF or B&Q) giving the systematic component, and on the other hand the structural component of physical capacity giving the dynamic aspect of capacity constraint reflected by equation system (3.7). There is an additional random component, given by the variance of CT, that realistically is always present. Based on equation system (3.7) and Fig. 3.6 we can enounce the following

Theorem of WIP (or Delay Theorem or Time-trap Theorem)

Given is a sequence of production steps with each process step having a different cycle time CT. Each process step with a longer CT than its preceding step is introducing a delay with the consequence of a potential increasing WIP formation in front of this process step. Such a process step is called time trap; therefore, a process may have more than one time trap.

Please note, the Theorem of Throughput is a direct consequence of this Time-trap Theorem; the bottleneck being simply the time-trap with the longest cycle time.

First Corollary to the Theorem of WIP (Corollary of Weak WIP Stationarity or Black Box Stationarity)

Necessary and sufficient condition that the WIP of a process presents a weak stationarity (total WIP remains constant), is that the exit rate ER of the process is equal to the input rate IR (e.g. takt rate TR), i.e. under ideal conditions the process will not build-up structural WIP. If the ER is greater than the TR, the process' WIP level will diminish, else increase. A process building-up WIP is called a constrained process. This corollary is valid for any transfer principle.

Lemma to the Theorem of WIP (Lemma of WIP Evenness)

If the cycle times CT_i of a process are all equal and show no variation (this is called a perfect balanced process), the WIP remains constant in front of each operation and depends only from the initial process conditions. In this case the WIP in front of each operation can be leveled out and reduced to the minimum, i.e. $\min WIP_i = 1$.

Second Corollary to the Theorem of WIP (Corollary of Strong WIP Stationarity or White Box Stationarity)

Necessary and sufficient condition to have a strong WIP stationarity of a manufacturing system, i.e. the WIP does not change at any time and at any stage of the process, is that the Corollary of Weak WIP Stationarity and the Lemma of WIP Evenness apply simultaneously.

We call it according to the usual mathematic interpretation weak stationarity, because only the average WIP remains time-invariant within the black box view; how it performs "white box" is not a topic of the first corollary, this is the topic of the second corollary. Please note, the two corollaries sound similar but are not the same. In fact, the second corollary is dealing with the internal stability of the WIP (balanced operations); indeed, the WIP will remain stable with the exception in front of the first operation (still in the backlog waiting buffer which will increase if the greater than zero sign of Eq. (3.9) applies). The First corollary is considering the whole production system even without considering the characteristic of the cycle times of operations. The balancing, i.e. the Lemma of Evenness, is not a necessary requirement for the Corollary of Weak Stationarity, being the First corollary self-sufficient giving the "black box" stationarity condition for the general process. The Corollary of Weak WIP Stationarity is one of the most important Corollary to be observed in order to manage WIP and therefore the capability of being able to supply the requested demand (see later Theorem of Lead Time Stability). The Corollary of Strong WIP Stationarity is the necessary and sufficient condition for an outperforming SPF JIT manufacturing such as transfer lines TFL represent. Equation (3.7b), however, determines in a "white box" consideration the spreading of the WIP along the line

Third Corollary to the Theorem of WIP (Corollary of WIP Spreading)

The WIP repartition in front of the operations along the production line depends on the difference of cycle times between two consecutive operations. The WIP will concentrate in front of an operation for heavily unbalanced cycle times or spread more evenly across the line if time traps are preceding the bottleneck.

This corollary states clearly that the bottleneck not necessarily has the highest WIP in front but depending whether time traps are preceding or not the bottleneck.

3.4 General Production Requirements for OTD Supply

A customer order is always defined by two attributes: the quantity and the expected delivery time EDT. If the quantity is a call-off, i.e. a fractioned repetitive quantity of an annual contract evenly distributed over the year, it can be advantageously translated into a takt rate TR. The customer expects that the desired quantity can be delivered, this is an aspect of production capacity, and he expects that the EDT can be matched, this is an aspect of MLT.

The necessary and sufficient conditions for an OTD, where OTD can be seen as the special case of the JIT philosophy at the end of the supply chain arriving to the external customer, can be expressed by the equation system (3.10) which has already been enounced in [5] comparing it to a LP optimization problem. Generally, as a direct consequence of the Theorem of WIP, two requirements have to be satisfied at the same time to have a customer-need viable production system: on the one hand, the specific capacity given by the ER_j of the smallest manufacturing cell C_j in a complex manufacturing system has to be greater than the customer's imposed TR, i.e. not showing capacity constraints complying to the Corollary of Weak WIP Stationarity, and on the other hand, the visible manufacturing lead time MLT_Z to the customer of the last process sequence Z of operations has to be shorter than the expected delivery time EDT, i.e. to be able to deliver on time. We will name equation system (3.10) the necessary and sufficient JIT **Takted Production Requirements (TPR) for an OTD supply**, i.e. for being able to satisfy customer demand. This is reflected by: first to have enough capacity (Eq. 3.10a) and second to be fast enough to be able to supply a manufacturing batch on time (Eq. 3.10b).

$$\begin{cases} \forall j : \inf\{ER_j\} \geq TR \\ Z : MLT_Z \leq EDT \end{cases} \qquad (3.10a, 3.10b)$$

Intrinsically, Lean production systems are based on paced SPF according to (D/D/1) queuing systems. More generally, we can generalize the TPR into the necessary and sufficient Fundamental **OTD General Production Requirements (GPR)** for generic production systems (G/G/1) independent of the transfer and manufacturing principles, by substituting the TR with a non-deterministic order rate OR (the greater equal sign has been replaced by the strict greater sign to avoid saturation related problems of the server) and adding the "visible" order backlog waiting time (BWT) to Eq. (3.10b) obtaining the following generally valid equation system (3.11)

$$\begin{cases} \forall j : \inf\{ER_j\} > E[OR] \\ Z : BWT_Z + MLT_Z \leq EDT \end{cases} \qquad (3.11a, 3.11b)$$

The BWT_Z is the backlog waiting time of a FIFO organized waiting queue in front of the entry process Z. If the (3.11a) is not satisfied, the BWT will increase according to the Corollary of Weak WIP Stationarity and (3.11b) can only be satisfied partially for certain orders through special scheduling prioritization, as e.g. short-term LIFO or priority FIFO, but it will not be possible generally for a FIFO scheduled order queue. Based on the equation system (3.10), or the GPR of equation system (3.11), we can therefore enounce the

Theorem of General Production Requirements (or OTD Theorem)

The necessary and sufficient conditions to supply a customer with OTD, i.e. with the right quantity at the right time, is that first the capacity requirement and second the lead time requirement have to be satisfied simultaneously, independent of the applied transfer principle, i.e. SPF or B&Q. The capacity requirement is given by the Corollary of Weak WIP Stationarity and the lead time requirement necessitates that MLT plus BWT is shorter than EDT.

The related key performance indicator KPI is the OTD, the on-time-delivery indicator, returning the percentage of time-conform deliveries to the customers. We will continue our reflection for the Lean-type manufacturing system. Notice, we have not put PLT in Eqs. (3.10b) or (3.11b) to allow on-time-delivery OTD as e.g. in [5, 8] but we have put here MLT, which is more appropriate, i.e. correct for the present theoretic considerations, because realistically not one piece but the whole ordered quantity (i.e. job equaling the ordinary batch) has to be delivered on time. This has not to be underestimated; indeed, more and more JIT deliveries have to be supplied also JIS, i.e. just-in-sequence, how the OEM is manufacturing the customized sequence, e.g. color of cars. For practical applications, i.e. for larger time frames of EDT and small batches, this might be of negligible importance; however, it is not for smaller time frames and large batches, especially not for a mixed-product cell and requires simultaneous delivery of several products.

In addition, within a regime of repetitive orders and with SPF principle, Eq. (3.10a) is predominant without neglecting (3.10b). Indeed, if Eq. (3.10a) is not satisfied, as a result WIP will increase making impossible to satisfy also Eq. (3.10b). Within a regime of traditional production environment based on B&Q principle, Eq. (3.10b) applies predominantly, having (3.10a) already incorporated with the Corollary of Weak WIP Stationarity. To generalize, the optimization problem of a production plant, given by the GPR equation system (3.11), can be simplified enouncing the following **Cardinal Objectives of a Production System (COPS)** formalized by the equation system (3.12), i.e. maximizing the throughput given by the ER and to speed-up processes to minimize PLT.[1]

$$\begin{cases} \max ER \\ \min PLT \end{cases} \tag{3.12}$$

Please note, equation system (3.12) is a general objective system not entering specifically into the flexibility of the line, i.e. mixed-product cellular manufacturing, for which we refer to Chap. 5 but for which it turns into a requirement. According to Little's Law expressed in Eq. (3.5), equation system (3.12) might be tautological at the first view but the interlinked variables optimize different objectives, and may address different production issues: capacity or speed as we will see in Sect. 6.5. Whereas it is always recommendable to minimize PLT, the

[1]ER and PLT representing the performance characteristic of a production system.

should be elastic and adaptable to the order load to have the most appropriate production capacity and therefore the lowest possible cost structure; overcapacity in capital-intensive industries is the present problem of European manufacturing plants, also amplified by economic cycles [9]. To overlook the whole value stream in integrated supply chains is therefore important in order to detect the real change of demand during business cycles [9]. The intrinsic dynamic of the repercussion of the demand along the value chain is leading to the pipeline-filling effect blowing-up demand artificially on all value-add levels of the chain [10]; this variation can partly be avoided by LM systems—but this is another topic.

After having now defined the COPS and associated laws, in the following, we will compare the performance of the two main transfer principles, i.e. B&Q and SPF (both push manufacturing principles) applied to generic pull system to comply to the objective system (3.12). We will compare it for an ideal (balanced) line and for the common case (non-balanced) as well as for mixed-product cell. We will also investigate how to react, i.e. parallelize or sequentialize the bottleneck for the instable case of a constraint when ER < TR, i.e. not satisfying the Corollary of Weak WIP Stationarity.

References and Selected Readings

1. Rother, M., Shook, J.: Learning to See, LEI/Cambridge Center, Cambridge (2003)
2. Rüttimann, B., Wegener, K.: Einführung in die Methoden von Lean Manufacturing und Six Sigma Quality Management, ETH Tools-IV Kurs, Lecturing notes HS2014, D-MAVT (2014)
3. Hopp, W., Spearman, M.: Factory Physics, International Edition. McGraw-Hill, New York (2000)
4. Curry, G., Feldman, R.: Manufacturing Systems Modeling and Analysis. Springer, Berlin (2011)
5. Rüttimann, B.G.: Discourse about linear programming and lean manufacturing: Two different approaches with a similar, converging rational. J. Serv. Sci. Manag. **8**, 85–91 (2015)
6. Goldratt, E.: The Goal, Excellence in Manufacturing. McGraw-Hill, New York (1984)
7. Goldratt, E.: Theory of Constraints. North River Press, New York (1999)
8. Rüttimann, B., Stöckli, M.: Going beyond triviality: The Toyota production system—lean manufacturing beyond Muda and Kaizen. J. Serv. Sci. Manag. **9**, 140–149 (2016)
9. Fischer, U., Rüttimann, B.: The Curse Of Globalization—Must We Expect Crisis in the Aluminium Industry that Are More Abrupt in the Future?, ALUMINIUM 85, 9. Giesel Verlag, Hannover (2009)
10. Rüttimann, B.: Dynamic of the Pipeline-Filling Effect, ALUMINIUM 5-2001. Giesel Verlag, Hannover (2001)

Chapter 4
Reducing Process Lead Time

After having introduced the three basic laws of production, given by the

- Theorem of Throughput
- Theorem of WIP, and the
- OTD Theorem (Theorem of General Production Requirements),

in the following sections we will develop the concepts of lead time performance and define its related theorems. We are at the first level of manufacturing complexity according to the TPS model of Fig. 2.3 dealing with the performance of a mono-product line. The performance is expressed by the lead times PLT and MLT as well as the transfer principles B&Q and SPF.

© Springer International Publishing AG 2018

B.G. Rüttimann, *Lean Compendium*, DOI 10.1007/978-3-319-58601-4_4

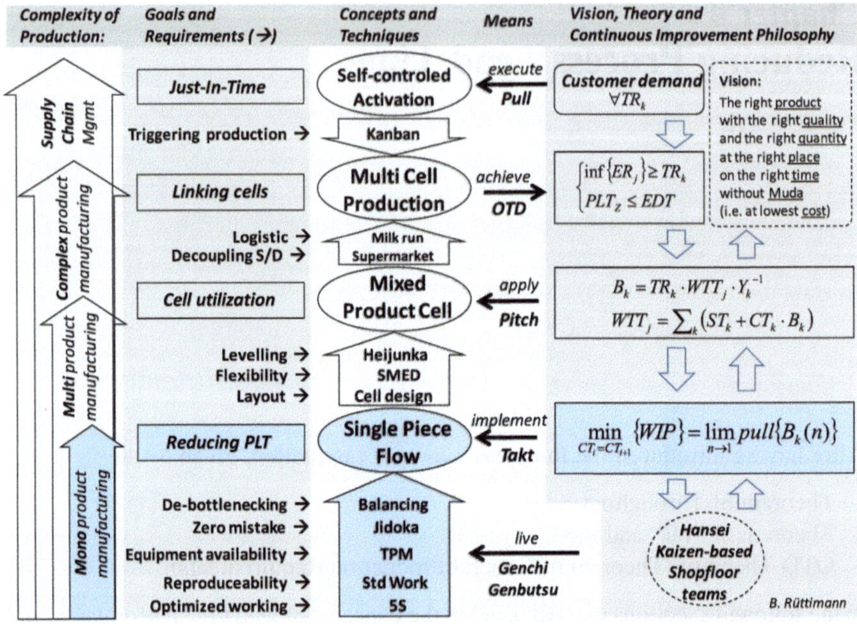

4.1 Performance of Different Transfer Principles in Balanced Lines

Let us take a fully balanced manufacturing line following the Lemma of WIP Evenness, i.e. the WIP is leveled to the minimum and remains constant in front of each operation for a SPF or is at the maximum B_k for B&Q. Fully balanced lines are typical in transfer lines (TFL) of the automotive industry or in mono-product or mixed-product manufacturing cells of various manufacturing and process industries (electronics, bottling). The characteristic of such a line or cell is defined in Eq. (4.1)

$$\forall i, k : CT_{i|k} = CT_{i+1|k} = CT_k \tag{4.1}$$

i.e. all operations have the same cycle time CT for a given product k, i.e. all elements of the column k of the matrix **A** have the same value. The ER is given by the inverse of the CT. The bottleneck, being all CT_i equal, is not defined. Usually, in the case of SPF in a mono, but also multi (mixed), product cell, as well as in a TFL, the first operation becomes the "drum" pushing the products through the manufacturing line, as we will see. Let us now compare such a balanced line of m operations by applying two different transfer principles for pieces, on the one hand a single piece flow SPF, and on the other hand a batch and queue B&Q principle with a batch size B_k for both cases (Fig. 4.1).

To the contrary of a B&Q principle, where the entire batch is machined before being moved from one operation to the other ($B_k(B_k)$ or "make-all, move-all"), within a SPF principle as soon as one piece has been processed it is moved to the next operation ($B_k(1)$ or "make-one, move-one"). Figure 4.1 shows exactly the instant of state transition when the machine cycle (or assembly operation) has finished and is ready to accept a new piece. The question is: how long does it take to manufacture the whole batch?

Let us now introduce the manufacturing lead time MLT, defined as the time when the first piece enters the line and the last piece of the batch leaves the last operation, in the case of a SPF, MLT can be calculated as follows for an empty line

$$MLT_{SPF} = (B_k - 1 + m) \cdot CT_k = \frac{B_k - 1 + m}{ER_k} \tag{4.2}$$

representing the **performance law of SPF manufacturing lead time** to calculate the lead time of a batch B_k to transit the balanced line (N.B.: balanced is factual identical to empty if $\sup\{WIP_i\} = 1$). In the case of $B_k = 1$ it gives apparently the same result as Little's Law, where m equals the WIP for a balanced SPF being independent of the batch size. For the B&Q principle, if no other batch is waiting, MLT can bc calculated as follows

$$MLT_{B\&Q} = B_k \cdot m \cdot CT_k = \frac{B_k \cdot m}{ER_k} \tag{4.3}$$

representing the **performance law of B&Q manufacturing lead time** giving with equal and deterministic CT the best achievable time for a batch B_k to transit the balanced line. In the case of several batches following, after transitory has expired, all operations have a $\sup\{WIP_i\}$ of size B_k in front (traditional B&Q with one batch queuing), $m*B_k$ corresponds to the total WIP in Little's Law. Equations (4.2) and (4.3) can be seen as the simplified (under ideal conditions) generalized Laws of Production Lead Time and represent the generalization of Little's Law, where Little's Law is just a case of the Law of Production Lead Time. Moreover, Eq. (4.4) also shows that Little's Law of PLT is a special case of the more generalized concept of MLT, being

$$\lim_{B_k \to 1} MLT(B_k) = PLT \tag{4.4}$$

When the batch size B_k becomes 1, Eqs. (4.2) and (4.3) are converging, as shown in Eq. (4.5), both transfer principles, i.e. SPF and B&Q, becoming undistinguishable, reflecting Little's Law, being in a non-discontinued manufacturing regime in front of each operation m one pieces, m representing then the WIP.

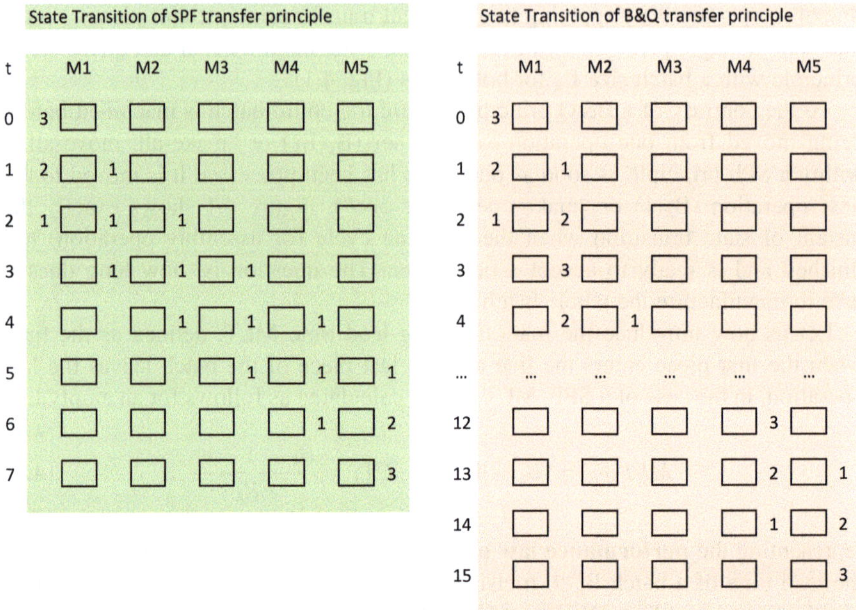

Fig. 4.1 Comparing MLT performance of a balanced line with CT = 1 for the two main transfer principles SPF and B&Q (schematically shown for one batch with size = 3, TR = ER, and m = 5 operations), with discontinued scheduling and no other queued batch to show the intrinsic difference between SPF and B&Q

$$\lim_{B_k \to 1} \{(B_k - 1 + m) \cdot CT_k\} = \lim_{B_k \to 1} \{B_k \cdot m \cdot CT_k\} = m \cdot CT_k = \frac{m}{ER} \approx \frac{WIP}{ER} \quad (4.5)$$

Please note, batch size 1 does not necessarily mean a one-off prototype but simply manufacturing one piece of a given product. Comparing Eq. (4.2) with Eq. (4.3) it shows, that $MLT_{B\&Q}$ will always be greater than MLT_{SPF} proving mathematically the known superiority of a SPF as summarized in Eq. (4.6).

$$MLT_{SPF} < MLT_{B\&Q} \quad (4.6)$$

This result has already been shown empirically in e.g. [1]. Please note, the ER is the same in both cases for SPF and B&Q, being the CT the same, which is not influenced by the manufacturing transfer principle.

Analyzing the sensitivity of MLT regarding batch size B_k, cycle time CT_i, as well as the number of operations m as shown in Fig. 4.2, the high sensitivity of B&Q is evident compared to SPF, e.g. in Eq. (4.7) shown for the number of operations m

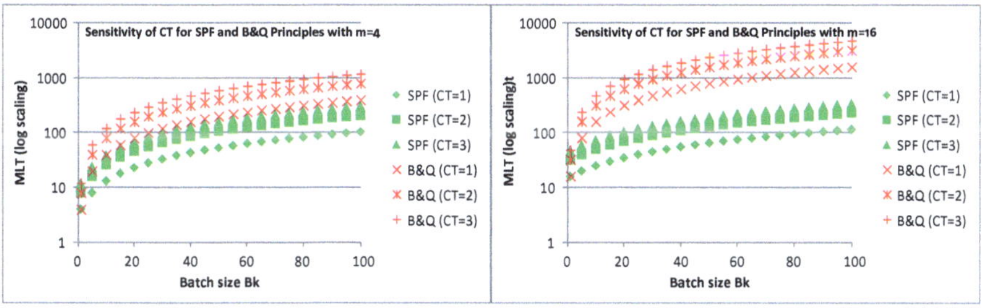

Fig. 4.2 Comparing production time in function of batch size B_k for the SPF and B&Q transfer principles of a balanced production line for different CT and m number of operations; high sensitivity of B&Q with regard to m

$$\frac{\partial}{\partial m} MLT_{B\&Q} = B_k \cdot CT >> \frac{\partial}{\partial m} MLT_{SPF} = CT \tag{4.7}$$

the B&Q having always a multiplicator in the equation which is the batch size. Therefore, products with many concatenated operations should be avoided to be manufactured with a B&Q modus as the evidence of Eq. (4.7) shows. Indeed, MLT of several months (!) are not unusual in certain manufacturing realities with several products k in the process and long queuing time; the same reasoning applies to the batch size. Indeed, batch size is the cause of the systematic generation of WIP. From Eq. (4.7) is perceivable the importance to reduce batch size to speed-up MLT to facilitate the OTD achievement; in fact, Eq. (3.10a) may be satisfied by both principles, but Eq. (3.10b) of lead time most probably will not. That is the main reason to reduce setup time (e.g. with the SMED technique) within a multi-product cell to reduce batch size.

In the case a flow with balanced CT can be implemented but the transfer unit is not one (e.g. $n = 2$, with the restriction $n < B_k$, resulting in a "two piece flow") the general case for n transferred units (with single cavity die) the Eq. (4.2) becomes

$$MLT_{nPF} = (B_k(n) - n + n \cdot m) \cdot CT_k \tag{4.8}$$

Equation (4.8) can be seen as the performance **Law of Generalized Lead Time** for a balanced line of a batch $B_k(n)$ for a product k and transfer principle n from which all other lead times can be derived for a balanced line with each operation m having equal CT. In the case of Eq. (3.1) being l an integer value, Eq. (4.8) can be written

$$MLT_{nPF} = (l \cdot n - n + n \cdot m) \cdot CT = (l - 1 + m) \cdot n \cdot CT$$

having again the structure of Eq. (4.2). The evidence that Eq. (4.8) is the generalized expression of lead time can be proven applying the limit conditions; indeed

$$\begin{aligned}
&\lim_{n\to 1}\{MLT_{nPF}\} = \lim_{n\to 1}\{(B_k(n) - n + n\cdot m)\cdot CT\} = MLT_{SPF}\\
&\lim_{n\to B_k}\{MLT_{nPF}\} = \lim_{n\to B_k}\{(B_k(n) - n + n\cdot m)\cdot CT\} = MLT_{B\&Q}\\
&\lim_{\substack{B_k\to 1\\ n\to 1}}\{MLT_{nPF}\} = \lim_{\substack{B_k\to 1\\ n\to 1}}\{(B_k(n) - n + n\cdot m)\cdot CT\} = PLT
\end{aligned} \qquad (4.9a - 4.9c)$$

Based on Eqs. (4.4), (4.5), and (4.9) we can enounce the following

Theorem of Generalized Lead Time (or Speed Theorem)

Given is a process with m cycle time CT-balanced operations; the manufacturing lead time MLT to produce a batch depends on the transfer principle, i.e. B&Q or SPF, or generally nPF, as well as on the batch size, determining the WIP of the process.

First Corollary to the Theorem of Generalized Lead Time (Corollary of Lead Time Convergence)

Given is a process with m cycle time CT-balanced operations. If the batch size tends to one, the two main transfer principles SPF and B&Q become undistinguishable, i.e. identical and the MLT converges at the limit to the PLT.

And also as a consequence of Eq. (3.6)

Second Corollary to the Generalized Lead Time (Corollary of Empty Line)

In the case that the line has no WIP in front of the operations, no delay is introduced and the PLT is equal to the sum of cycle times CT corresponding to the work content WC.

Lemma to the Generalized Lead Time (Lemma of SPF Desirability)

The superior performance of SPF versus B&Q leads to the natural interest to always try to implement a SPF to maximize production performance.

Balanced SPF production lines encounter a big interest in high volume manufacturing lines because of superior performance compared to, usually unbalanced, B&Q systems (see already tayloristic Ford T-model production in the 1920s); in addition they offer also simplified calculations. A stable, i.e. steady-state, process is essential. But it has also to be stated, the application of SPF is not always possible and therefore the B&Q principle will retain its domain of application. As we have seen, balanced CT is a necessary but not sufficient requirement for strong WIP stationary processes. Necessary and sufficient **Requirements for a perfect Lead Time Stable Process** are given by the equation system (4.10)

$$\begin{cases} ER > TR \\ CT_i = CT_{i+1} \end{cases} \qquad (4.10a, 4.10b)$$

which is a consequence of the Corollary of Strong WIP Stationarity. In the ideal case of $\sup\{WIP_i\} = 1$, the second condition assures that no "Muda" in form of waiting time (delaying constraint) is generated within the process and the first condition assures that no structural WIP (capacity constraint) is build-up, warranting that the PLT will remain stable. If only Eq. (4.10a) is observed, the process remains stable but less performant than when also Eq. (4.10b) is satisfied. As a consequence of the Law of WIP summarized with equation system (3.7) and based on equation system (4.10) we can enounce the

Theorem of Lead Time Stability (or Steady State Theorem)

Necessary and sufficient condition that a manufacturing process has an overall (black box) constant lead time is, that the Corollary of Weak WIP Stationarity applies.

Corollary to the Theorem of Lead Time Stability (Corollary of Strong Lead Time Stability)

Necessary and sufficient condition that a manufacturing process has at any time, at any stage (white box) a constant lead time, i.e. a uniform flow, is that the Corollary of Strong WIP Stationarity applies.

In the case the PLT remains stable we can link MLT to PLT, i.e. combining PLT Eq. (3.5) of Little with the MLT Eq. (4.2) and we get the **MLT-PLT Conversion Law for the SPF transfer principle**

$$MLT_{SPF} = (B_k - 1) \cdot CT_i + PLT_{SPF} \tag{4.11}$$

Analog to Eq. (4.11) we can derive the **MLT-PLT Conversion Law for the B&Q transfer principle** by defining Eq. (4.12)

$$MLT_{B\&Q} = B_k \cdot PLT_{SPF} \tag{4.12}$$

Equations (4.11) and (4.12) are cardinal equations which reflect an important property of balanced production lines linking the process lead time PLT of a single piece to the manufacturing lead time MLT to produce the entire batch with the SPF principle or the B&Q principle. This fact is a consequence of equation system (4.9) representing the Law of Generalized Lead Time (4.8).

The importance to calculate MLT is more pronounced for production systems based on B&Q transfer principle than SPF based systems, because usually SPF based systems are often applying Lean concepts with reduced batch size. Indeed, as we have seen from Eq. (4.9c) the difference between MLT and PLT shrinks. Nevertheless, MLT is also important for Heijunka leveled large manufacturing cells or long transfer lines, to seize the batch for the Heijunka box timeslot, as we will see.

4.2 Performance of Different Transfer Principles in Unbalanced Lines

Let us derive the formulas of Eqs. (4.2) and (4.3) for an unbalanced line, i.e. with different CT according to the regime of Eq. (4.13) still for a mono-product manufacturing cell, i.e. with the elements of the column vector of matrix **A** for the product k now being different. We will derive the generalized equations for the case

$$\forall i, k : CT_{i|k} \neq CT_{i+1|k} \tag{4.13}$$

It has to be pinpointed right from the beginning, that the Corollary of Strong Lead Time Stability will never apply due to the not applicability of the Lemma of WIP Evenness. Further, if the Corollary of Weak WIP Stationarity cannot apply, i.e. not being satisfied, the system will "explode" if TR > ER with increasing WIP and unstable PLT never being able to achieve the requirements of the equation system (3.10) and neither to comply to the GPR (3.11). Usually, a SPF is only applied in the case of a balanced line, but a flow can theoretically also be implemented for an unbalanced line, of course with inefficiencies (Muda) between the operations given by machine or piece waiting times leading to a delay in MLT according to the Law of WIP described in equation system (3.7); of course, the CT differences have to be small. Based on empirical deduction, to Eq. (4.2) has to be added a WIP generated delaying term leading to Eq. (4.14),

$$MLT_{SPF} \approx (B_k - 1 + m) \cdot \left[\inf CT_{i|k} + \sum_{i=1}^{m} (CT_{i|k} - \inf CT_{i|k}) \right] + \sum_{i=1}^{m} \sup \beta_{k|i}(t) \cdot (CT_{i|k} - \inf CT_{i|k})$$

CT-balanced term *CT non-balanced term* *delaying term* (4.14)

decomposing the lead time into three terms, where $\inf\{CT_{i|k}\}$ is the shortest cycle time in the line and $\beta_{k|i}(t)$ is the portion of the WIP from the batch B_k in front of the m_i operations. Please note, $\beta_{k|i}(t)$ is generated dynamically in front of each operation m_i with $CT_i > CT_{i-1}$ and, as a consequence, depending from the sequence of unbalanced CT_i, Eq. (4.14) is therefore only a conceptual approximation which decomposition into three distinctive parts are: a CT-balanced term, a CT non-balanced term, and a delaying term where $\sup\{\beta_{k|i}(t)\}$ is the maximum observed WIP. Eq. (4.14) may underestimate the MLT. These are typical simulation problems of discrete programming type where the algebraic formulation of WIP dynamic is difficult to be represented and therefore approximate; simulations are better suited to see the performance (see also Fig. 4.3 and Sect. 4.5 The effects of stochastic CT and OR variability on performance).

Please note, according to Eq. (3.7b) the bottleneck has not necessarily the greatest WIP as usually divulged. The size of the WIP depends from the difference of two consecutive CT, i.e.

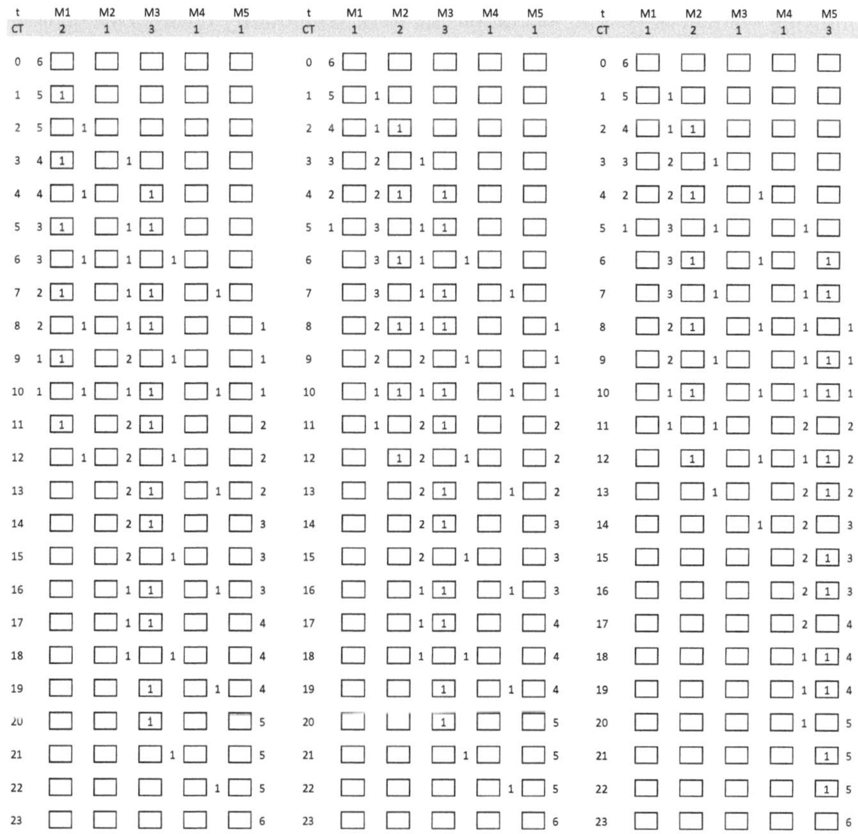

Fig. 4.3 Didactic examples of discrete simulations of MLT_{SPF} and WIP development in unbalanced production lines with production constraint, i.e. TR > ER (illustrative case with empty line and discontinued production after the batch has been processed)

$$dWIP_i = \left(\frac{1}{CT_{i-1}} - \frac{1}{CT_i}\right)dt$$

and

$$WIP_{i|\tau_2} = \int_{\tau_1}^{\tau_2} \left(\frac{1}{CT_{i-1}} - \frac{1}{CT_i}\right) \cdot dt + WIP_{i|\tau_1} = \frac{\frac{CT_i}{CT_{i-1}} - 1}{CT_i} \cdot t\Big|_{\tau_1}^{\tau_2} + WIP_{i|\tau_1}$$

shows, that if $CT_i > CT_{i-1}$ the WIP_i will increase not showing the property of strong stationarity, and depends from the characteristics of initial condition. In fact, it depends as well as from the preceding time traps, limiting intra-process departure rate DR_{i-1} and therefore AR_i with $AR_i < TR$ having therefore less WIP in front of a following bottleneck (risking operations m_i to be temporarily idled). This demonstrates that the common belief that the bottleneck presents the biggest WIP is wrong, the WIP of the bottleneck depending from the preceding time traps (see also Fig. 4.3 and figures of Sect. 4.5). The bottleneck still remains unequivocally defined as the longest CT in a line.

Similarly to SPF suiting better a balanced line, a B&Q is typically applied in the case of an unbalanced line and the $MLT_{B\&Q}$ is

$$MLT_{B\&Q} \approx B_k \cdot \sum_m CT_{m|k} + \sum_m \sup\beta_{k|i}(t) \cdot \left(CT_{i|k} - \inf CT_{i|k}\right) \tag{4.15}$$

If the line is empty, the delaying term is not present. From Eq. (4.15) it can be easily derived, that Eq. (4.3) is a special case of Eq. (4.15) when all CT_i are equal; indeed, different than SPF, Eq. (4.3) can be calculated precisely also in the case of an unbalanced line becoming Eq. (4.16)

$$\lim_{CT_i = CT_{i+1}} \left(B_k \cdot \sum_{i=1}^{m} CT_{i|k} + \sum_{i=1}^{m} \sup\beta_{k|i}(t) \cdot \left(CT_{i|k} - \inf CT_{i|k}\right) \right) = B_k \cdot m \cdot CT_{i|k} \tag{4.16}$$

The analog conclusion to Eq. (4.16) of the B&Q principle is also valid for the SPF principle, being Eq. (4.2) a special case of Eq. (4.14) by observing Eq. (4.17). Notice, to obtain Eq. (4.2) from Eq. (4.17) the cycle times have to be balanced leading to a new optimal $opt\{CT_i\}$ and/or to split $CT_i > \inf\{CT_i\}$ into additional operations with finally a new number M of operations, where $M^* < M$ being $opt\{CT_i\} > \inf\{CT_i\}$

$$\begin{aligned} \lim_{CT_i \to \inf CT_i} \left[(B_k - 1) \cdot CT_i + PLT_{SPF} \right] &\approx (B_k - 1 + M) \cdot \inf\{CT_i\} \\ &= (B_k - 1 + M^*) \cdot opt\{CT_i\} \end{aligned} \tag{4.17}$$

A simplified presentation, such as in Fig. 4.1, is only possible for discrete cycle times which are multiple of the time step, therefore it is not possible to show for machine cycle time CT having continuous data characteristics not matching the time step; for that a more sophisticated simulation package is necessary, or we have to deal with fractional products (liquids).

In Fig. 4.3 the case is shown for three empty production lines each of it has five steps and the same sum of cycle times, i.e. the same work content, but with the operations distributed differently; the batch size of 6 pieces is the same. To fabricate the batch in an empty line, the MLT_{SPF} is 23 time units, independent of the bottleneck's localization; it shows clearly, that the bottleneck with cycle time $CT_b = 3$ time units limits the ER of the whole process for all three cases.

$ER_b - \frac{1}{CT_b} \approx \lim\limits_{t \to T} \frac{\int_0^t ER \cdot d\tau}{t} = ER_{process}$ i.e.e.g $ER_b = \frac{1}{3} \approx \frac{\int_{t1}^{t2} ER \cdot d\tau}{t_2 - t_1} = \frac{6 \ 1}{23-8} = \frac{5}{15}$ reflecting the Theorem of Throughput after transitory. In all three cases the first piece exits at instant 8 from the line. This leads to enouncing the

Third Corollary to the Theorem of Throughput (Corollary of Bottleneck Location)
Regarding process lead time for an empty line, it is indifferent where the bottleneck is located, whether at the begin, or in the middle, or at the end of the process; the manufacturing lead time MLT is the same depending only from the selected transfer principle.

The WIP develops differently according to where the time traps are located. Although we have an unbalanced line, SPF has been applied; this is possible because the batch is limited to $B_k = 6$ (e.g. according to a Heijunka Box scheduling of a mixed-product cell). In reality if the batch would be greater and continue to have a TR = 1, the WIP would increase according to equation system (3.7) the production line not being able to process enough pieces with two constraints. Moreover, even if the batch would remain to be 6, i.e. afterwards another batch B_{k+1} would continue to enter the production line, showing for a SPF an unstable situation. In the case of a B&Q transfer principle would be applied to the exercise of Fig. 4.3, the calculated MLT is according to Eq. (4.15)

$$MLT_{B\&Q} = 6 \cdot (2 + 1 + 3 + 1 + 1) = 48$$

$MLT_{B\&Q} = 48$, too long to be shown in Fig. 4.3. The same consideration about TR > ER applies to the B&Q principles. To derive a universal valid Law of Generalized Lead Time according to Eq. (4.8) for unbalanced cycle times we use a paradigmatic approach and not the reasoning of Eq. (4.14). In the case of a non-balanced line, Eq. (4.8) turns into Eq. (4.18) becoming the manufacturing lead time for unbalanced empty lines

$$MLT_{nPF}^{g} = (B_k(n) - n) \cdot \sup\{CT_i\} + n \cdot \sum_{i=1}^{m} CT_i \tag{4.18}$$

and by setting in Eq. (4.18) $CT_i = CT_{i+1}$ Eq. (4.18) turns back into Eq. (4.8) the special case of a balanced line.

Further, to apply proficiently a SPF having the same work content of the case in Fig. 4.3, a balanced line with $CT_{opt} = 1.6$ time units is necessary and according to Eqs. (4.2) or (4.18)

$$MLT_{SPF} = (6 - 1 + 5) \cdot 1.6 = 16$$

the MLT_{SPF} results then to be 16 time units only, less than the 23 in the unbalanced case. To show the universal character of Eq. (4.18) let us proof that Eq. (4.8) is a special case of (4.18), indeed

$$\lim_{CT_i=CT_{i+1}} \left\{(B_k(n)-n)\cdot \sup\{CT_i\} + n\cdot \sum\nolimits_i CT_i\right\} = (B_k(n)-n)\cdot CT + n\cdot m\cdot CT$$

With this evidence let us enounce the

Main Theorem of Production Time (or SPF Dominance Theorem)
Regardless of the characteristic of a manufacturing system, i.e. with balanced or with unbalanced cycle times CT, the manufacturing lead time MLT is always shorter for a single piece flow SPF transfer principle than for a batch and queue B&Q transfer principle.

Corollary to the Main Theorem of Production Time (Corollary of Lead Time Limit)
In the case of a CT balanced line, the SPF principle presents the shortest achievable MLT and therefore also the shortest achievable PLT.

Therefore, as it has just been proven, and this might be heretic to the present beliefs, we can state the following

Lemma to the Main Theorem of Production Time (Lemma of SPF Regime)
It is always recommendable to implement a SPF principle also for unbalanced production lines; although some equipment may be waiting in a SPF, there is no lost capacity, because the bottleneck is always fully loaded and the ER remains the same but the MLT is shortened.

The problem consists therefore not primarily in how to balance the operations in order to have the favorable conditions of the Corollary to the Main Theorem of Production Time (Corollary of Lead Time Limit) to minimize MLT but primarily how to exploit this theorem following the Lemma of SPF Regime by implementing the layout to allow a one piece flow. The next step is to try to balance it, and in the case TR > ER how to debottleneck the production line. Next, we will discuss this last topic. How to flow we will see in Sect. 4.4 and how to conceive a cell in Sect. 5.3.

Equation (4.18) is valid for an empty line; this is rather rarely the case. Usually, in classic B&Q job shop manufacturing systems, the production line is not empty but in front of the operations are products waiting to be machined (or assembled, whatever). Equation (4.18) has therefore to be completed with the queuing time WIP/ER_b according to Little's law. Equation (4.18) becomes therefore

$$MLT^G_{nPF} = (B_k(n)-n)\cdot \sup\{CT_i\} + n\cdot \sum_{i=1}^{m} CT_i + \sum_{i=1}^{b} WIP_{i|k-1}\cdot \sup\{CT_i\}$$

or

$$MLT_{nPF}^{G} = n \cdot \sum_{i=1}^{m} CT_i + \sup\{CT_i\} \cdot \left(B_k(n) - n + \sum_{i=1}^{h} WIP_{i|k-1} \right) \quad (4.19)$$

At this point we want to pinpoint and remark that Little's Law is only correct for a queue in front of a single workstation, i.e. not for a cell and not for a line composed of several operations. This can be clearly deducted from Eq. (4.19) which represents the most comprehensive law of lead time. Equation (4.19) can be considered to be the **Universal Performance Law of Generalized Lead Time for Non-balanced Lines** where the term $\sup\{CT_i\}$ corresponds to the bottleneck with ER_b and the WIP extends to all operations in front of the bottleneck. Notice, the amount of WIP is subject to the instant the new batch $B_k(n)$ experiences a delay. Further, Eq. (4.19) reveals an important insight leading to the following

Lemma to the Theorem of Generalized Lead Time (Lemma of Push Manufacturing Principle)

In push manufacturing principles, to shorten manufacturing lead time MLT of a batch entering the line, it is advisable to have the bottleneck at the begin of the line; this is valid independent of the transfer principle.

Notice, this does not contrast the third Corollary to the Theorem of Throughput (Corollary of Bottleneck location), being valid for empty lines. The Lemma of Push Manufacturing Principle is linked to the

Lemma to the Theorem of General Production Requirements (Lemma of Flexible Scheduling Principle)

To keep until the latest instant the maximum flexibility in scheduling to allow preferential order treatment bypassing general FIFO principle of backlog, e.g. to observe OTD for a single important order, it is advisable to release the next order into shopfloor only when the first operation is imminent ready to accept it.

The Lemma of push manufacturing principle avoids filling the production line and avoids queuing on the shopfloor congesting the line. It is better to concentrate it into the backlog, increasing BWT which allows a flexible scheduling principle. The Lemma of flexible scheduling principle reduces further MLT, releasing the production order JIT when the first operation is ready to accept a new batch; this allows to having the maximum flexibility of the order backlog buffer, changing priority of batches and allows to bypassing FIFO scheduling principle at the last moment. The application of both lemmas is advisable in the case a SPF cannot be implemented. The Lemma of Flexible Scheduling Principle will lead to the Lemma of "Input Equals Exit" Principle in Sect. 7.1 Generalized Kanban technique. The insight gained by Eq. (4.19) leads to state further

Lemma to the Theorem of Throughput (Lemma of Throughput Control)
Also for a balanced line it is advisable to create an artificial deterministic bottleneck to control better the line, i.e. to reduce the randomness of its performance due to stochastic influences on CT.

To understand the importance of this lemma, think of a Chinese shopfloor with dozens of persons cutting, trimming, and sewing jeans. It is hardly imaginable to overlook every single operator at his workplace; creating an artificial bottleneck, it is sufficient to control the bottleneck operation to be sure to maintain OTD also with a B&Q transfer principle.

4.3 Debottlenecking: Parallelization or Sequentialization?

To have a perfect stable process performance, i.e. the PLT of a piece results equal for all pieces of the batch reflected by the Theorem of Lead Time Stability (Steady state Theorem), a necessary but not sufficient condition is that the ER has to be greater than the TR or generally E[OR], i.e. the Corollary of Weak WIP Stationarity applies. The additional necessary and sufficient condition for a strong PLT stability is that the "white box" WIP_i remains constant, i.e. the Corollary of Strong WIP Stationarity applies leading to the Corollary of Strong Lead Time Stability; to do so, the line has also to be balanced. The necessary and sufficient conditions can be described therefore by generalizing equation system (4.10) enouncing the necessary and sufficient **Requirements of General PLT Stability (long-term steady state) for a Manufacturing System** with the Eq. (4.20)

$$\{PLT = const : |ER > E[OR], WIP = const\} \tag{4.20}$$

The condition WIP = const seems to over specifying Eq. (4.20) but this is not the case; indeed, if WIP is not constant, also PLT will not be constant. In the case ER < E[OR], PLT will increase leading to an undesirable unstable process performance with the consequence of not being compliant with the OTD requirements of equation system (3.11). This can be avoided by limiting the order rate, i.e. reducing TR, but this is not a practically viable solution. The solution consists of debottlenecking, i.e. to reduce the CT_b of the bottleneck which, in the case of Fig. 4.4, is also a constraint for the general production requirement of equation system (3.11). Please note, the bottleneck is only a serious issue when it is also a constraint, i.e. with $CT_b > TT$; if $CT_b < TT$ it is only the operation with the longest CT which may be used as "drum" to implement a paced SPF.

Debottlenecking has to be made by reducing the work content of the bottleneck operation, as well as all other constraints, in two ways: (a) redistribute, i.e. spread excessive time to other process steps according to Eq. (4.17), or (b) redistribute to a new additional operation; we will discuss further alternative (b). Alternative (b) can be performed again in two ways: (b1) by parallelization, i.e. to duplicate the

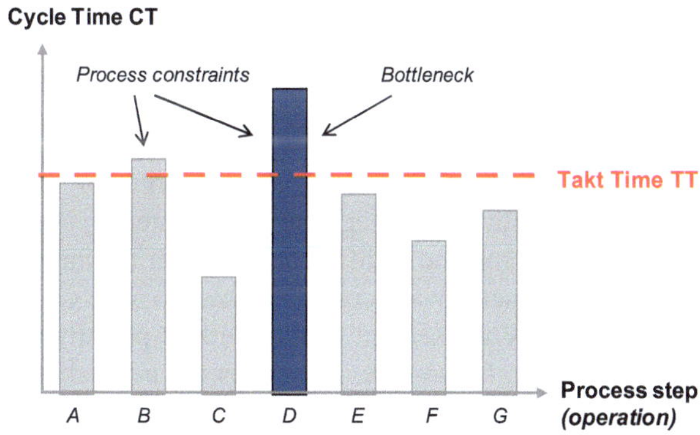

Fig. 4.4 Time-operation chart of an unbalanced manufacturing cell or transfer line from [2]

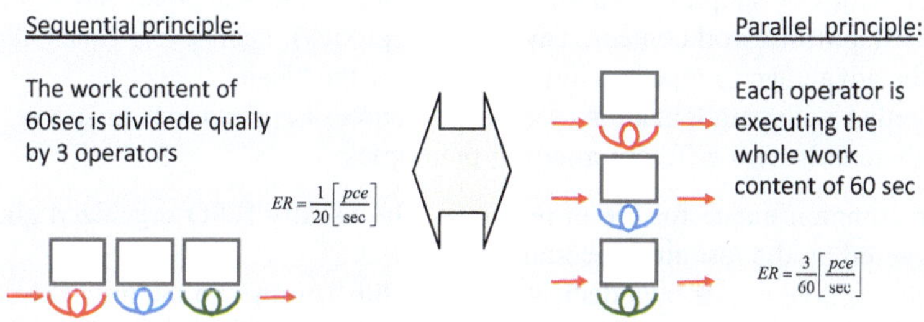

Fig. 4.5 Debottleneck principles: parallelization or sequentialization

resource of the bottleneck, or (b2) by sequentialization, i.e. by fractioning the work content in two consecutive operations. Which one is better?

In Fig. 4.5 are shown both concepts. Although mathematically there is no difference in capacity performing both at the same ER; pay attention, intrinsically 1/20 is not the same as 3/60 which leads to the paradoxon of the mean. This is only the same if we assume an even distribution. Indeed, the exits in the parallel case may not be (if not organized) distributed evenly over the 60 s. However, parallelization is easier to implement and to be adapted to required capacity changes. However, sequentialization allows better to industrialize and complies to the concept of flow; in addition, the whole equipment has not to be doubled and made available in excess. For the layout and staffing we refer to Sect. 5.3.

Summarizing, we can define the three **debottlenecking principles** as follows:

– spreading, i.e. repartition of the exceeding unbalanced time of the bottleneck to other operations; this suits best, if the exceeding CT is relatively small compared to the other operations and can even be neglected if $CT_b < TT$; in this case we have an unbalanced line and the bottleneck should become the pacemaker. This is different if the excess is considerable; if that is the case and spreading is not viable there are existing two alternatives

– sequentialization, i.e. cutting the bottleneck task into two, i.e. an additional new operation, or more fractional operations, to attain required TT
– parallelization, i.e. doubling, or tripling, the bottleneck resource with identical workstations.

The debottleneck principles are a direct interpretation of the more general **organization principles** (splitted or integrated workcontent), see Sect. 5.3. Usually when consulting, I recommend a different approach for industry and service companies (of course, approach always contingent also to other circumstances); the main reason is intrinsic how the service or the work is performed. Indeed, apart from the content of the task, for service companies the parallelization (integrated workcontent) to duplicate the resource is easy to implement, e.g. for call center operators or cashier in supermarket, not necessitating big investments; this might not be the case in industry. It has also another advantage linked to the functional characteristics of office workplaces for which we refer to [3].The procedural characteristics of shopfloor workplaces recommend the sequentialized solution if possible (splitting workcontent, Tayloristic approach); short cycle times allow to have the advantage to repeat activities and favor the "Kata".

Together with parallelization of similar resources (workstations or server, whatever) there exist two different **queuing principles**:

– one common queue for the m resources; the usually FIFO organized queue is triggered by the resource becoming free
– each resource has its own dedicated queue; this is rather suitable for the case of specific non interchangeable resources. Keep in mind, the queue has not to be FIFO organized but can be managed according to one of the scheduling principles.

Notice, the manufacturing lead time for a batch, i.e. a quantity of identical products, shows no difference between parallelization and sequentialization within a general queuing system (G/G/s); however is queuing time added, how the queuing is organized may influence the performance of the system, i.e. with one common queue for the s servers or with s queues, i.e. one queue for each server, i.e. machine.

By the way, joking sarcastically we could state that parallelization is the problem of European overcapacity, sequentialization has the same effect to debottlenecking as parallelization, with reduced PLT, in addition it favors the correct pacing to obtain a customer takted line (or Panta Rei as Greek philosopher Heraklit said) with all benefits of Kata.

Theorem of Debottlenecking (or Dual Solution Theorem)

In presence of a general time trap (and even more for the bottleneck), and in presence of a constant inter arrival rate AR, it is indifferent from an output metric performance DR, if the time trap is doubled or the work content of the time trap is fractionized and sequentialized.

First Lemma to the Theorem of Debottlenecking (Lemma of Adaptability)
If the inter arrival times AR are not constant, the parallelization has to be preferred over sequentialization allowing, if technically possible, a better adaptation of the time trap to the changing arrival situation.

Rather than for a workstation, which would mean double investment, this is recommendable for a manufacturing cell, where the capacity can be increased or decreased by adding or reducing operators within a circuit-organized cell (see Sect. 5.3).

Second Lemma to the Theorem of Debottlenecking (Lemma of High Performance)
To assure Kata in high performance manufacturing line, the sequentialization has to be preferred over parallelization.

Third Lemma to the Theorem of Debottlenecking (Lemma of Ticketing or Queue Sharing)
In the case of parallelization of the time trap, and in presence of not constant processing times, it is recommendable to have a single queue for several servers by pooling variability of process times to shorten average backlog waiting time BWT.

Fourth Lemma to the Theorem of Debottlenecking (Lemma of Urgency)
Not only in the mandatory case of a constrained system, impacting profitability, but also generally, all the attention should be directed to the bottleneck, because it limits directly the exit rate ER of the process according to the Theorem of Throughput.

4.4 Creating Flow

As it clearly emerges from the mono-pillar TPS model of Fig. 2.2, the aim of the TPS is not to install Kanban-controlled inventory management, as often falsely believed, but a SPF. Apart from the increased process performance of a SPF compared to B&Q in terms of PLT, according to the Main Theorem of Production Time and the relative corollary as well as lemma on which we put the focuses in this book, Toyota creates flow wherever possible to improve the process. By the way, the concept of flow is not a Toyota invention but based on Taylorism and has been applied first in Ford's T-model production. Interesting to mention, as Jones from the LEI (Lean Enterprise Institute) investigated, already the Venetian Arsenal, the largest industrial site of the world in the eighteenth century, used concepts of standardized work and reduced transportation ways, implementing a sort of flow on the final assembling to build their ships. This clearly shows that certain manufacturing techniques are necessary to maximize production output and are intrinsic of the production system to optimize performance also with regard to PLT.

A flow entails several advantages. Indeed, lowering WIP not only PLT becomes faster and decreases invested capital improving liquidity, but a SPF let emerge problems of product quality and all other types of inefficiencies or Muda, such as downtime due to setups or breakdowns, which a high WIP would hide. We will here not enter into the difficulties to install a flow, with potential setbacks because of line standstill due to lack of pieces in the line (e.g. quality or inbound logistics issues), leading sometimes to abandon again the SPF in favor of the B&Q; for the necessary perseverance you can consult the classic Lean books. We will drive here the attention more into the conceptual aspects of flow.

However, before entering the topic how to implement a flow, let us define what a flow is. As the word suggests, it has the characteristics of a fluid but with discrete material physics, i.e. more precise a SPF or also called a one piece flow. To have a laminar, i.e. uniform, and a stationary, i.e. time invariant, flow, any sequential particle must have the same dynamics property (best represented by the concise concept of vectorial divergence $\mathrm{div}\underline{q} = 0$). For our manufacturing system this means to have a balanced line, i.e. each operation has the same CT for a given product k. When the transferred units n becomes one and is moved in a takted pace from one station to the next, the manufacturing principles push and pull become indistinguishable. We can therefore define a SPF with Eq. (4.21)

$$SPF := \left\{ \lim_{n \to 1} Push\{B_k(n)\} = \lim_{n \to 1} Pull\{B_k(n)\} | CT_i = CT_{i+1} \right\} \quad (4.21)$$

and consequently we can now even enounce the

Central Limit Theorem of Manufacturing CLTM (or Definition of Perfect SPF or Manufacturing Principle Identity Theorem)

In the case of a balanced manufacturing line with equal cycle times CT at each work station, when the transferred unit tends to one, the two manufacturing principles Push and Pull become indistinguishable, defining a perfect balanced SPF.

Corollary to the Central Limit Theorem of Manufacturing (Corollary of Improper SPF)

If the transferred quantity tends to one, but the cycle times CT are not equal (unbalanced line), this condition defines an improper SPF (or SP handling); the manufacturing principle push or pull may still be distinguished and depends from the triggering.

Equation (4.21) and this CLTM now define unequivocally what a SPF is, beyond the approximation of the taxonomy matrix of Fig. 3.2. The consequence of this theorem is profound; indeed, it gives evidence that the latent perception that the taxonomy of Fig. 3.2 is not dogmatic and even less axiomatic, and demonstrates that SPF is the limit of two different manufacturing principles. Therefore, the manufacturing principles push and pull are rather to be referred to the triggering of the production whereas SPF and B&Q to the transfer principle (this is the reason why we call it transfer principle instead of manufacturing modus). This sounds to be in contradiction to the concept of “drum”, i.e. push, but rather critical for the

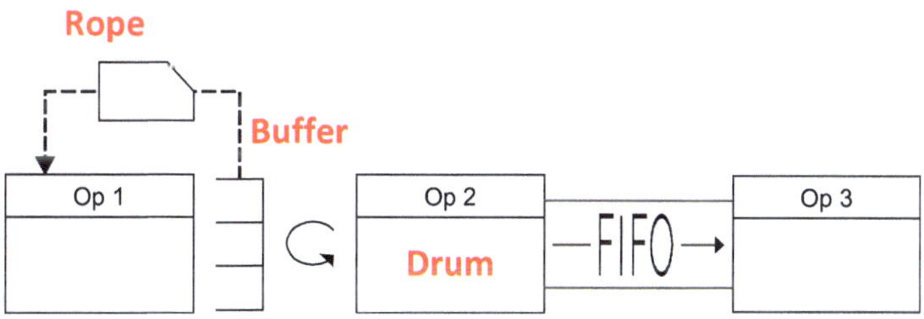

Fig. 4.6 Goldratt's "Drum-Buffer-Rope" technique to create and maintain flow (push-pull)

definition is where the entry point of the pull signal is. Notice, we are now entering the topic of sophism; "flow on pull" is therefore the sufficient and perfect naming for this manufacturing modus operandi.

Flow can be implemented mainly within two entities (see also Fig. 4.6):

- transfer line: create a takted, also called paced, transfer line for cumbersome products, e.g. typically automotive assembly lines
- manufacturing cell: create a takted manufacturing cell with usually U-shaped or Z-shaped layout for small pieces, e.g. electro-technical or mechanical components

How to layout and staff a manufacturing cell will be dealt with in Sect. 5.3. Here we deal with the conceptual mechanism of controlling the triggering of production. Due to the Theorem of Throughput and the Lemma of Urgency the first step is to identify the bottleneck. Then, in a simple, non-complex manufacturing entity the Drum-Buffer-Rope (DBR) technique is applied to the bottleneck. The DBR approach is a direct consequence of Goldratt's theory of constraints TOC [4, 5]. For complex manufacturing entities, later we will see variations of the DBR technique, not necessarily applied to the bottleneck but also to link manufacturing cells. The DBR as shown in Fig. 4.6 helps to figuratively explain the concept. Such as in a Roman galley the bottleneck becomes the drum beating the general takt according its intrinsic CT, imposing the pace to the downstream flow. All the other operations have to follow; no overproduction in the downstream operations will materialize because being able to process immediately the arriving parts (see next Sect. 4.5 Eq. (4.22)), beside of existing time traps. The "drum" has always to be loaded and never run out of work to maximize profit, being the ER directly linked to the CT_b of the bottleneck (Theorem of Throughput). The "buffer" has therefore always to be filled in order never to run out of stock and is replenished as soon as the trigger level is reached.

Such buffers are strategic buffers, necessary to keep production running creating a regime of push-pull. To have the buffer always filled, it is recommended to have short CT upstream of the pacemaker. This is the evidence, that the zero stock aim is an illusion and represents only the minimizing objective with local optima. Further, the downstream operations following the "drum" with the cycle times CT_i have to be ideally balanced, i.e. they have to show equal cycle times. To analyze the operations, the time-operation chart of Fig. 4.4 can be applied and the Dual Solution Theorem may help to debottleneck in the case of a constraint. The DBR technique is the evidence, that a flow generally is considered to be a single piece push transfer principle, which complies to the taxonomy of Fig. 3.2. However intrinsically, the CLTM applies.

Ideal situation is to have a balanced FIFO line which allows to transferring the single piece in a takted manner with the FIFO buffer to be minimized; this is a contingent optimization depending on the variability of CT and the operational reserve in the case of an operation downtime. If cycle times are not balanced, according to the Corollary of Improper SPF it may also be possible to implement a SPF, but rather a pull SPF. Then it is the downstream operation triggering the production of the upstream workstation representing the pure modus operandi of the taxonomy shown in Fig. 3.2 (see also Chap. 7). It may be recommended to have an in-line Kanban to separate the consecutive workstations (see Sect. 7.2). To balance the line a time-operation or time-operator chart such as shown in Fig. 4.4 is recommended. The takt, or the Kanban have to be strictly observed; no overproduction is allowed because creating potentially a WIP representing Muda and delaying throughput.

4.5 The Effects of Stochastic CT and OR Variability on Performance

Until now we have modeled the multi-step manufacturing line with a deterministic CT of Dirac's δ-function best represented by queuing systems (D/D/1). An unbalanced line may also be the consequence of a balanced line with $E[CT_i] = E[CT_{i+1}]$ but where the CT show variation; we can call this a stochastically balanced line opposed to perfectly balanced line. In this section, we will generalize the reasoning and analyze the effect of CT_i variability to PLT, i.e. a queuing system of type (D/G/1) and after with OR respectively IR-derived variability of type (G/G/1), (M/G/1), and finally (M/M/1), the last one giving an insight how fully flexible Industry 4.0 manufacturing systems may potentially behave. Indeed, a general server process with CT distribution of normal type is better suited than a Markov-type, the latter having implicitly a greater range of variation than normal type distribution with controllable standard deviation. Markovian exponential density function might therefore not be the appropriate distribution to model server cycle time, a normal distribution being more suitable to model real TFL. However, if the underlying Poissonian requirements are fulfilled, the usage can be justified for arrival rate, such as order rate OR, less for input rate IR, because the IR into the system has to show low variability and can be managed by purpose. Nevertheless, the Markov distribution is of particular interest in queuing theory due to its simple calculation, as showed in Sect. 3.3. The arrival rate entering the first operation, being according to our definition the input rate IR into the manufacturing system, is scheduled, and therefore can be considered to remain deterministic D without variation or general G normal type with very low variability, which is also the case for the TR in a limited time period (i.e. non-sensitive to e.g. seasonality).

Given these introductory explications to justify a (D/G/1) queuing system, let us at first assume to have an ideal simplified manufacturing line, ideal for computational reasons and simplified to understand variability, with CT_i having the following properties

$$\forall i : \sup\{E[CT_i] \geq E[CT_{i+1}]\} := CT \tag{4.22}$$

which, according to Eq. (3.7b) as well as the Corollary of Empty Line, allows to assume an empty production line which has no deterministic structural WIP in front of each operation except the first one. This assumption is theoretically possible because the CT through parallelization or sequentialization, or generally by the help of the debottlenecking principles, can always be constructed to satisfy Eq. (4.22). Strictly seen, if we leave the domain of deterministic type, to avoid WIP formation, Eq. (4.22) is only true if the following condition applies

$$\forall i : \inf\left\{E[CT_i] \pm Z_{\alpha/2} \cdot \sqrt{\frac{Var(CT_i)}{n}}\right\} > \sup\left\{E[CT_{i+1}] \pm Z_{\alpha/2} \cdot \sqrt{\frac{Var(CT_{i+1})}{n}}\right\}$$

where $Z_{\alpha/2}$ stands for the standardized normal distribution, and α is the complement to the significance level, very hard to be realistic. Further strictly seen, the variability of CT impeeds the observation of Eq. (4.22) with the consequence, that WIP will form in front of a time trap if the stochastic variability of $CT_i < CT_{i+1}$ is given, as we will see. For simplicity and comprehension reason we will proceeding our reasoning based on Eq. (4.22) with the equal sign, getting as a result the best achievable case under ideal conditions. Further, if the CTs are no more deterministic but, as in reality they show variation, let us now continue to assume that the CT_i are random variables and present a normal distribution

$$\forall i : N(E[CT_i], Var[CT_i])$$

The normal distribution can be assumed, if $E[CT_i]$ is quite far from the zero with a limited variance in order that the natural inferior process limit is greater than zero; otherwise a lognormal distribution has to be taken, complicating slightly the comprehension. Based on the Corollary of Empty Line, the expected mean of the process lead time E[PLT], without WIP due to Eq. (4.22), is given by Eq. (4.23), where m denotes the number of operation steps or workstations

$$E\left[PLT_{empty}\right] \approx \sum_{i=1}^{m} E[CT_i] = m \cdot CT = WC \tag{4.23}$$

however, what will be the variance Var(PLT) of the process lead time? From statistical theory, we can expect that the Central Limit Theorem of statistics applies getting

$$SD\left(E\left[PLT_{empty}\right]\right) = m \cdot \frac{SD(CT)}{\sqrt{m}} \tag{4.24}$$

To confirm the variance, we will use Monte Carlo simulation for a process consisting of five operations m = 5, each following the distribution of N(1; 0.1). In Fig. 4.7 we see the simulation results of a batch with batch size $B_k(1) = 300$ confirming Eqs. (4.23) and (4.24): it is interesting to note, that the coefficient of variation Cv = 0.23/5 of the sum of the five single operations CT is smaller than that of the individual operations with Cv = 0.1/1 being a direct consequence of the Central Limit Theorem.

If we assume again the same five operation, but with $E[CT_i] = 5$ for all i, in order to apply a normal distribution without interfering the lower bound with its natural limit zero, but with different variances according to Fig. 4.8 represented on the left side, the simulation of the workcontent WC is shown in Fig. 4.8 on the right side. We talk here about WC and not PLT because high variation forms WIP between the operation, not being empty any more, although we will not always be so strictly in the naming, implicitly intending the meaning from context. In this case Eq. (4.24) becomes

$$SD(E[WC]) \approx \sum_{i=1}^{m} \frac{SD(CT_i)}{\sqrt{m}} \tag{4.25}$$

Figure 4.8 on the right side confirms Eq. (4.25).

Let us now simulate the resulting work content WC variation of a five-step line with each operation m_i having an exponential Markov-distribution of cycle times with λ = 1, i.e.

$$\forall i : M\left(E[CT_i] = \frac{1}{\lambda}, Var[CT_i] = \frac{1}{\lambda^2}\right)$$

Those type of distributions are expressly avoided in Lean type manufacturing lines through e.g. TPM (total productive maintenance) to limit variation of the workstation but will most probably be realistic in Industry 4.0 shopfloor environment being the direct consequence of fully flexible workstations addressable by AGVs (automated guided vehicles); whereas in Lean the variance might be equipment or operator related, in Industry 4.0 it will be rather customer and variable product mix imposed. Indeed, Industry 4.0 has the objective to manufacture a wide customizable product range specified through the internet of things IOT [6, 7]. Due to the fact of the presence of a natural lower bound and due to unforeseeable delays, each operation will in Industry 4.0 environment rather present a right skewed CT

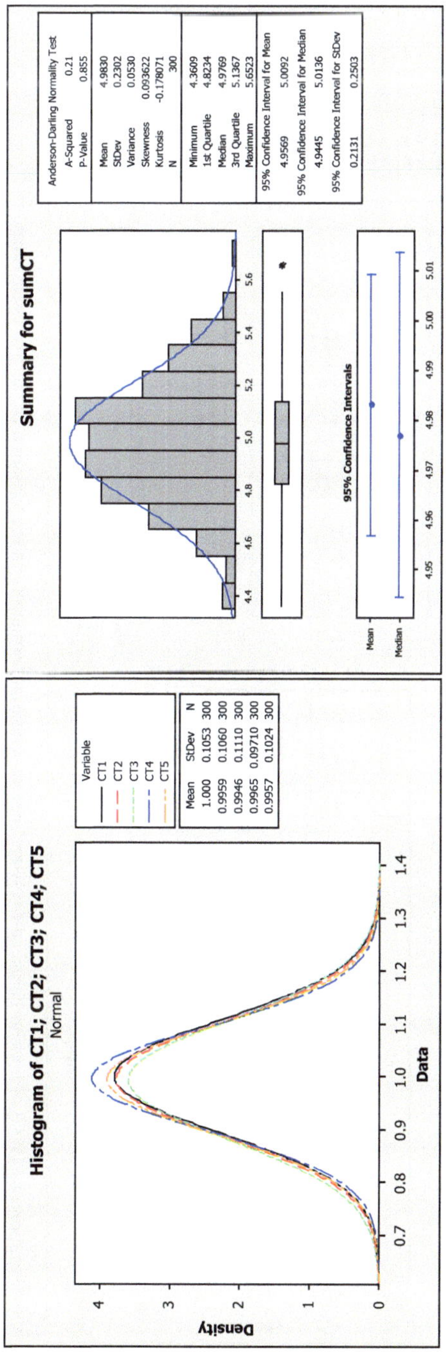

Fig. 4.7 CT density distributions of the five operations and resulting histogram of Monte Carlo simulated PLT for the sequence of the five operations, each operation m_i with normal-type N(1; 0.1) distributed CT_i

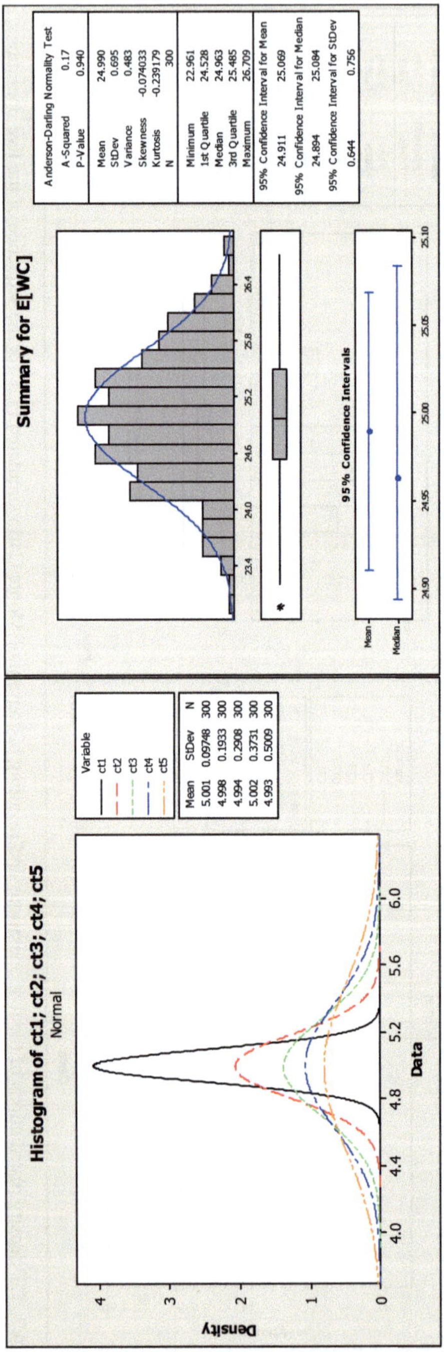

Fig. 4.8 Variation of E[WC] of a mean-balanced five-step line but with high variability in the variation of the five cycle times

distribution, very well represented by an exponential distribution. Indeed, the lower bound of $M(1/\lambda; 1/\lambda^2)$ being zero would mean, that the operation with $CT_i = 0$ is a skipped operation. Notice, we prefer to discuss the variability of work content WC in terms of Markovian CT dimension rather than of Poissonian AR and DR; indeed, the measures are usually calculated in time units (e.g. with MTM (methods, time, measurements), standard work), as it represents better the practical manufacturing reality. Instead of PLT we prefer to talk here of WC because the PLT in such an unbalanced line due to the stochastic variation in the distribution will have forcedly waiting time, never being able to observe condition (4.22) mentioned above.

As we can expect, also for this CT distribution apply Eqs. (4.23) and (4.24), the sum of the cycle times giving the E[WC] = 5. The variation of the WC however is much larger than in the previous case of normally distributed cycle times, being SD (E[sumCT]) = 2.2 due to the fact that for a single CT with Markov distribution E[CT] = 1/ λ and also SD(CT) = $1/\lambda$, by the way, that is what makes this distribution so easy for manual calculus. The resulting Cv factor of the sum of cycle times is 2.2/5 smaller as the Cv of single cycle times but compared to the normal distribution much greater. From Fig. 4.9 it emerges that 25% of pieces have a resulting WC exceeding 6.12, i.e. 1.12 time units above the mean of 5 causing problems of OTD but also for in-house JIT logistics. The fact that 25% of pieces are below 3.27 time units does not change the situation; indeed in automotive OEM an anticipated delivery might also not be desired, the workstation or cell not being able to accept the delivery, the buffer not yet being empty. All this shows which deleterious repercussion variation of CT has.

As per Figs. 4.7 and 4.9, the simulation confirms both hypothesized Eqs. (4.23) and (4.24). This lead to the

Theorem of Stochastic Cycle Time Variability (or Convergent Workcontent Theorem)

Given is a process with non-deterministic distributed cycle times CT each with the same mean and variance, i.e. a stochastically balanced line. The mean and the variance of the sum of the cycle times, i.e. the time to execute the work content WC, follows the Law of Central Limit Theorem of Statistics multiplied by the number of operations.

Despite of 300 simulation cycles, the resulting WC of five stations shown in Fig. 4.9 does not have normal distribution shape. This high variation of the cycle times is therefore not apt for a performant manufacturing system as we will show in the simulation on the next pages. If we change the concept of WC to PLT, or even more MLT, for a 300 piece batch the situation even gets worse, as we will see.

Let us now try to simulate a process with m = 5 operations, each operation having the same E[CT] = 1 minute per piece, i.e. a theoretically balanced process suiting ideally for SPF, considering also the input rate IR and the WIP influences on process performance. Further, let the CT have a "limited" variability of SD (CT) = 0.1 due to random performance loss of the equipment but also compensated or improved work execution of the operator, the $E[CT_i]$ remains 1 min, it represents in fact a stochastically balanced line. We will simulate at first a push manufacturing

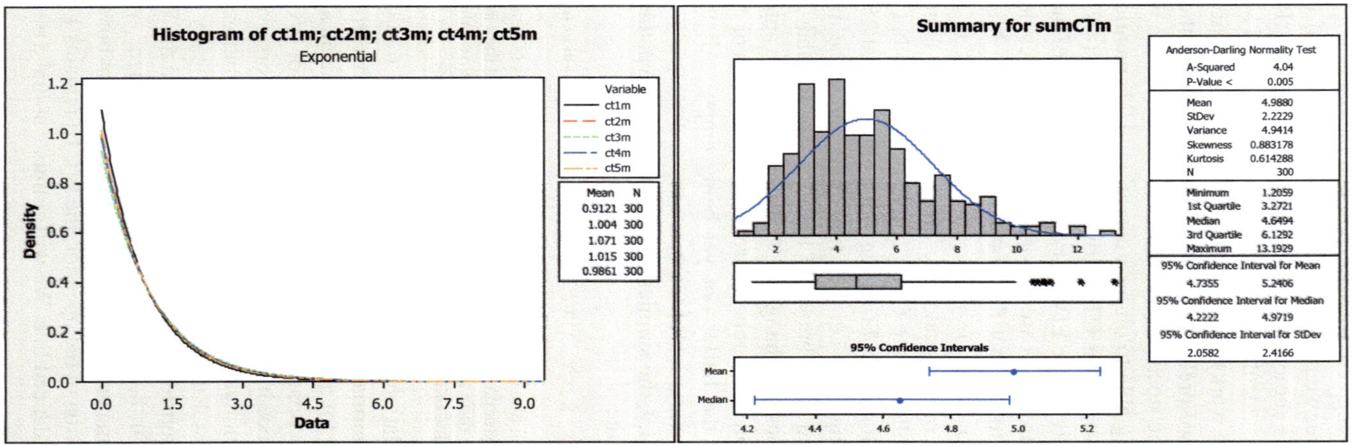

Fig. 4.9 CT density distributions of the five operations and resulting histogram of Monte Carlo simulated variation of WC for the sequence of the five operations, each operation m_i with Markovian exponential-type distributed CT_i with $\lambda = 1$

principle with IR = 1 piece/min, complying theoretically to the stochastically balanced CT of the line represented by a (D/G/1) queuing system. Due to the Law of WIP according to Eq. (3.7a–c) and its Theorem of WIP with relative Corollaries, the random generated unevenness of the five operations CT due to CT variability will generate WIP. Due to the difficulty to represent discrete manufacturing events with non-deterministic CT within discrete time intervals, as already encountered in Fig. 4.3, let us rely again on hydraulic modeling, not talking about parts (pieces) but rather liquids (liter). We have a constant inflow with IR = 1 liter per minute; the operation with CT is substituted by an outflow tube with random capacity and $E[ER_i] = 1$ liter per minute. Let us apply the following schematic algorithm to simulate the WIP evolution

```
begin
WIPi(t=0):=0 {no queue at the beginning}
FGI(t=0):=0 {the final reservoir is empty at the beginning}
for t=1...300 do {let us simulate 300 cycles}
    for i=1...5 do {compute the WIP for the 5 stations}
        if i=1
        then if WIP1(t−1) + IR(t)·Δt ≥ ER1(t)·Δt
             then exit1(t) := ER1(t)·Δt {according to the capacity}
             else exit1(t) := WIP1(t−1) + IR(t)·Δt {according to the
level of reservoir}
             WIP1(t) := WIP1(t−1) + IR(t)·Δt − exit1(t)
        else if WIPi(t−1) + ERi−1(t)·Δt ≥ ERi(t)·Δt
             then exiti(t) := ERi(t)·Δt
             else exiti(t) := WIPi(t−1) + ERi−1(t)·Δt
             WIPi(t) := WIPi(t−1) + ERi−1(t)·Δt − exiti(t)
    end
    FGI(t) := FGI(t−1) + exit5(t)
end
stop
```

An extract of the first 50 runs from the tabular output of the simulation for a (D/G/1) queuing system is shown in Fig. 4.10.

The system at the base is that of Fig. 3.6 and the algorithm executes at any time cycle of 1 min the liquid to advance by one operation in the quantity of the variable pipe section (capacity) available. Figures 4.11, 4.12, 4.13, 4.14, and 4.15 show the simulation for a process with five operations regarding WIP forming in front of each operation, the IR evolution and distribution, total WIP in the process, Exit rate at the end of the process cycle time, ER at the bottleneck. PLT follows approximate the evolution of total WIP due to the fact that calculated according to Little's Law taking the smallest value for the ER with different IR characteristics. The total WIP is therefore our key indicator.

The result of the simulation is interesting and reflects the enounced Theorems of Throughput and WIP and its corollaries. In the above simulations we have a

t	IR(t)	WIP1(t)	ER1(t)	exit1(t)	WIP2(t)	ER2(t)	exit2(t)	WIP3(t)	ER3(t)	exit3(t)	WIP4(t)	ER4(t)	exit4(t)	WIP5(t)	ER5(t)	exit5(t)	FGI(t)	ERb(t)	WIPtot(t)
0		0			0			0			0			0			0		0
1	1	0	1.03865433	1	0	0.96564399	0	0	1.0211039	0	0	0.9726439	0	0	0.93674479	0	0	0.93674479	0
2	1	0	1.1075813	1	0	1.07502623	1	0	1.06204661	0	0	0.95222489	0	0	1.09473825	0	0	0.95222489	0
3	1	0	1.04975652	1	0.0289378	0.9710622	0.9710622	0.09596956	0.87509265	0.87509265	0	1.12497732	0	0	0.95836203	0	0	0.87509265	0.12490735
4	1	0.0110964	0.9889036	0.9889036	0	1.01951175	1.01784139	0	1.23596958	1.11381095	0.20696561	0.90684534	0.90684534	0	1.22378405	0	0	0.90684534	0.21806201
5	1	0.0845442	0.9265522	0.9265522	0	1.21485338	0.9265522	0	1.09790531	0.9265522	0.27588418	0.85763363	0.85763363	0	1.05528318	0.85763363	0.85763363	0.85763363	0.36042838
6	1	0.03732266	1.04722154	1.04722154	0.13229209	0.91492945	0.91492945	0	1.08851038	0.91492945	0.23884893	0.9519647	0.9519647	0	1.1282572	0.9519647	1.80959833	0.91492945	0.40846368
7	1	0.23450689	0.80281577	0.80281577	0.01283649	0.92227137	0.92227137	0	0.99044128	0.92227137	0.18347708	0.97764322	0.97764322	0.01804999	0.95959324	0.95959324	2.76919157	0.80281577	0.44887044
8	1	0.34893903	0.88556786	0.88556786	0	1.07472436	0.89840435	0	1.21156162	0.89840435	0.02155653	1.0603249	1.0603249	0.0403453	1.03802959	1.03802959	3.80722115	0.88556786	0.41084086
9	1	0.30054296	1.04839607	1.04839607	0	1.23482016	1.04839607	0.03845482	1.00994125	1.00994125	0	1.0523135	1.03149778	0.12478614	0.94705694	0.94705694	4.75427809	0.94705694	0.46378392
10	1	0.37776673	0.92277623	0.92277623	0.02703177	0.89574446	0.89574446	0	0.95651435	0.93419928	0.08719377	0.8470055	0.8470055	0	1.03173007	0.97179165	5.72606973	0.8470055	0.49199227
11	1	0.53493429	0.84283244	0.84283244	0	1.09893839	0.86986421	0	1.06218672	0.86986421	0	1.22925508	0.95705799	0	1.14989607	0.95705799	6.68312772	0.84283244	0.53493429
12	1	0.57123593	0.96369836	0.96369836	0	1.21263645	0.96369836	0	1.15416568	0.96369836	0.03038351	0.93331485	0.93331485	0	1.04848673	0.93331485	7.61644257	0.93331485	0.60161944
13	1	0.74671384	0.8245221	0.8245221	0	1.04780202	0.8245221	0.02256518	0.80195691	0.80195691	0	0.91675063	0.83234042	0	0.98047954	0.83234042	8.44878299	0.80195691	0.76927902
14	1	0.79106871	0.95564513	0.95564513	0.02141515	0.93422998	0.93422998	0.01195723	0.94483794	0.94483794	0	1.09928326	0.94483794	0	0.94905641	0.94483794	9.39362093	0.93422998	0.82444108
15	1	0.77282959	1.01823911	1.01823911	0.13152966	0.9081246	0.9081246	0	1.03950647	0.92008183	0	0.95067352	0.92008183	0.02399884	0.89608298	0.89608298	10.2897039	0.89608298	0.9283581
16	1	0.75393379	1.0188958	1.0188958	0.2109569	0.93946856	0.93946856	0.01007732	0.92939124	0.92939124	0	0.96205879	0.92939124	0	1.08306641	0.95339009	11.243094	0.92939124	0.97496802
17	1	0.74451859	1.0094152	1.0094152	0.24455555	0.97581656	0.97581656	0	1.03023807	0.98589388	0.06897732	0.91691656	0.91691656	0	1.14005901	0.91691656	12.1600105	0.91691656	1.05805146
18	1	0.6839135	1.06060509	1.06060509	0.20019557	1.10496507	1.10496507	0.0907483	1.01421676	1.01421676	0.27954785	0.80364623	0.80364623	0	1.00030418	0.80364623	12.9636568	0.80364623	1.25440523
19	1	0.66249334	1.02142016	1.02142016	0.3722671	0.84934863	0.84934863	0.09868427	0.84141266	0.84141266	0.21200335	0.90895716	0.90895716	0	1.03056121	0.90895716	13.8726139	0.84141266	1.34544807
20	1	0.68895849	0.97353485	0.97353485	0.40475502	0.94104693	0.94104693	0.03247836	1.00725285	1.00725285	0.29754793	0.92170827	0.92170827	0	0.98946738	0.92170827	14.7943222	0.92170827	1.4237398
21	1	0.65909384	1.02986465	1.02986465	0.42158554	1.01303414	1.01303414	0	1.09249942	1.04551249	0.32511166	1.01794876	1.01794876	0	1.05627102	1.01794876	15.812271	1.01303414	1.40579103
22	1	0.74855812	0.91053571	0.91053571	0.27613922	1.05598203	1.05598203	0.1795909	0.87639113	0.87639113	0.15780702	1.04369577	1.04369577	0.05524387	0.9884519	0.9884519	16.8007229	0.87639113	1.41733914
23	1	0.65022367	1.09833446	1.09833446	0.4573322	0.91714148	0.91714148	0.15610804	0.94062434	0.94062434	0.07589391	1.02253745	1.02253745	0.18158382	0.8961975	0.8961975	17.6969204	0.8961975	1.52114164
24	1	0.70931574	0.94090793	0.94090793	0.48781884	0.91042129	0.91042129	0.04950066	1.01702866	1.01702866	0.06070748	1.0322151	1.0322151	0.09280925	1.12098967	1.12098967	18.81791	0.91042129	1.40015197
25	1	0.69849497	1.01082077	1.01082077	0.56988957	0.92875005	0.92875005	0	1.10476208	0.97825071	0.06134877	0.97760942	0.97760942	0.22122647	0.8491922	0.8491922	19.6671022	0.8491922	1.55095977
26	1	0.62502839	1.07346658	1.07346658	0.63445654	1.0088996	1.0088996	0	1.10657377	1.0088996	0.17190396	0.89834441	0.89834441	0.19146727	0.92810361	0.92810361	20.5952059	0.89834441	1.62285616
27	1	0.6302952	0.99473319	0.99473319	0.65661139	0.97257834	0.97257834	0	1.1045162	0.97257834	0.21521623	0.92926607	0.92926607	0.19977793	0.92095541	0.92095541	21.5161613	0.92095541	1.70190075
28	1	0.66238184	0.96791336	0.96791336	0.69900461	0.92552015	0.92552015	0.09011171	0.83540844	0.83540844	0.08532668	0.96529798	0.96529798	0.18691289	0.97816302	0.97816302	22.4943243	0.83540844	1.72373773
29	1	0.81263837	0.84974348	0.84974348	0.50541063	1.04333745	1.04333745	0.24552461	0.88792455	0.88792455	0	1.17904999	0.97325123	0.11717924	1.04298488	1.04298488	23.5373092	0.84974348	1.68075285
30	1	0.89630612	0.91633225	0.91633225	0.54535539	0.87638748	0.87638748	0	1.19863971	1.12191209	0.18222097	0.93969112	0.93969112	0.06593658	0.99093378	0.99093378	24.5282429	0.87638748	1.68981907
31	1	0.99253343	0.90377269	0.90377269	0.50790557	0.94122251	0.94122251	0	0.96215869	0.94122251	0.13118486	0.99225862	0.99225862	0.07582887	0.98236633	0.98236633	25.5106093	0.90377269	1.70745273
32	1	1.0745461	0.91798733	0.91798733	0.54375935	0.88213356	0.88213356	0	0.91580377	0.88213356	0.10858447	0.90473395	0.90473395	0	1.19899751	0.98056282	26.4911721	0.88213356	1.72688991
33	1	1.23738038	0.83716572	0.83716572	0.36611191	1.01481316	1.01481316	0	1.11644169	1.01481316	0	1.13803471	1.12339762	0	1.16218648	1.12339762	27.6145697	0.83716572	1.60349229
34	1	1.03168018	1.2057002	1.2057002	0.50496694	1.06684517	1.06684517	0	1.1219599	1.06684517	0.01159882	1.05524635	1.05524635	0.02487194	1.03037441	1.03037441	28.6449441	1.03037441	1.57311788
35	1	1.14195518	0.88972501	0.88972501	0.33644722	1.05824472	1.05824472	0.06798754	0.99025718	0.99025718	0.00198408	0.99987192	0.99987192	0.09315848	0.93158538	0.93158538	29.5765295	0.88972501	1.6415325
36	1	1.00998203	1.13197315	1.13197315	0.47519124	0.99322914	0.99322914	0.14638824	0.91482844	0.91482844	0	1.02143519	0.91681252	0	1.1347142	1.009971	30.5865005	0.91482844	1.6315615
37	1	1.06920962	0.9407724	0.9407724	0.251065	1.16489864	1.16489864	0.34415137	0.96713551	0.96713551	0	0.98662408	0.96713551	0	1.00794098	0.96713551	31.553636	0.9407724	1.66442599
38	1	1.02250209	1.04670753	1.04670753	0.0523993	1.24537322	1.24537322	0.60446289	0.9850617	0.9850617	0	1.29790888	0.9850617	0.02869343	0.95636827	0.95636827	32.5100043	0.95636827	1.70805772
39	1	1.12757548	0.89492662	0.89492662	0.19895958	0.74836634	0.74836634	0.3647548	0.98807443	0.98807443	0	1.0669248	0.98807443	0	1.26396233	1.01676787	33.5267722	0.74836634	1.69128985
40	1	1.22009716	0.90747832	0.90747832	0.12169065	0.98474725	0.98474725	0.47260548	0.87689657	0.87689657	0	1.30688489	0.87689657	0	1.28159043	0.87689657	34.4036687	0.87689657	1.81439329
41	1	1.26288026	0.9572169	0.9572169	0	1.1324681	1.07890754	0.33684765	1.21466537	1.21466537	0.09883316	1.11583221	1.11583221	0.02540893	1.09042328	1.09042328	35.494092	0.9572169	1.72397001
42	1	1.39864621	0.86423405	0.86423405	0	0.92059842	0.86423405	0.44100967	0.76007203	0.76007203	0	1.13049169	0.8589052	0	0.94450071	0.88431413	36.3784061	0.76007203	1.83965588
43	1	1.32983054	1.06881567	1.06881567	0.10768698	0.9611287	0.9611287	0.44174993	0.96038844	0.96038844	0	1.07241721	0.96038844	0	0.97612877	0.96038844	37.3387946	0.96038844	1.87926744
44	1	1.4081227	0.92170784	0.92170784	0	1.10757811	1.02939481	0.36412117	1.10702357	1.10702357	0.22601675	0.88100683	0.88100683	0	0.97916429	0.88100683	38.2198014	0.88100683	1.99826062
45	1	1.4552096	0.9529131	0.9529131	0	0.96918224	0.9529131	0.33997719	0.97705708	0.97705708	0.16130201	1.04177181	1.04177181	0.14709776	0.89467405	0.89467405	39.1144754	0.89467405	2.10358656
46	1	1.45002835	1.00518125	1.00518125	0	1.07646659	1.00518125	0.28036988	1.06478855	1.06478855	0.24505305	0.98103752	0.98103752	0.08326102	1.04487426	1.04487426	40.1593497	0.98103752	2.05871231
47	1	1.35256195	1.0974664	1.0974664	0.00504719	1.09241922	1.09241922	0.37897615	0.99381295	0.99381295	0.23781847	1.00104753	1.00104753	0.01957112	1.06473743	1.06473743	41.2240871	0.99381295	1.99397488
48	1	1.46559361	0.88696834	0.88696834	0.0484954	0.84352013	0.84352013	0.02577148	1.1967248	1.1967248	0.54844577	0.8860975	0.8860975	0	0.96401254	0.90566862	42.1297558	0.84352013	2.08830626
49	1	1.60575232	0.85984129	0.85984129	0	0.97550291	0.90833669	0	1.11789493	0.93410817	0.54066341	0.94189052	0.94189052	0	1.0112773	0.94189052	43.0716463	0.85984129	2.14641573
50	1	1.52056294	1.08518938	1.08518938	0.14395736	0.94123202	0.94123202	0	0.98119771	0.94123202	0.47163601	1.01025943	1.01025943	0.05819447	0.95206496	0.95206496	44.0237112	0.94123202	2.19435078

Fig. 4.10 Output of the Monte Carlo simulation for the first 50 cycles for a D/G/1 system with IR: D = 1 and ER_i: G = N(1; 0.1)

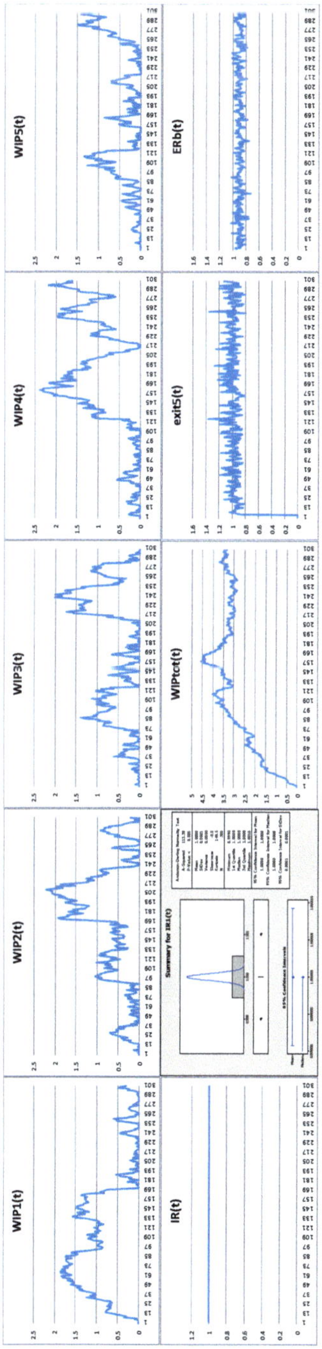

Fig. 4.11 Monte Carlo simulation of WIP formation due to variation in CT and IR with deterministic IR = 1/min and stochastic CT with E[CT] = 1, SD [CT] = 0.1 simulating a (D/G/1) queuing system

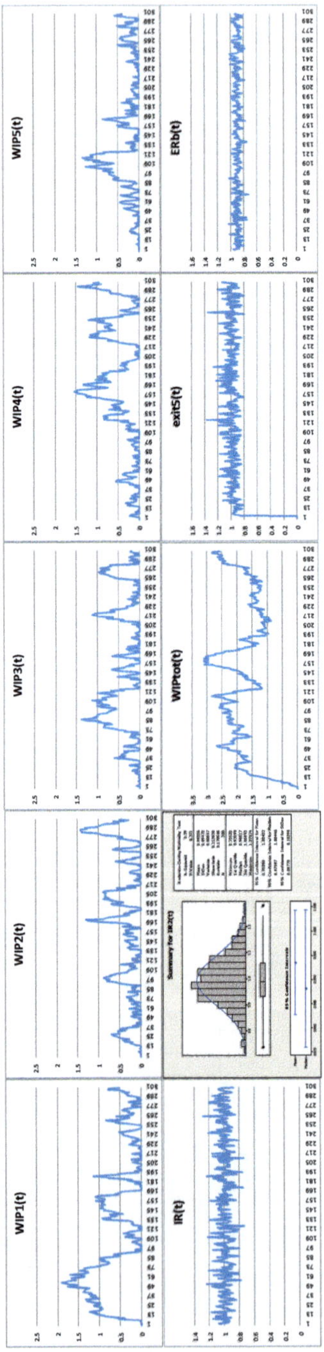

Fig. 4.12 Monte Carlo simulation of WIP formation due to variation in CT and IR with E[IR] = 1/min, SD[IR] = 0.1; E[CT] = 1, SD[CT] = 0.1 simulating a (G/G/1) queuing system

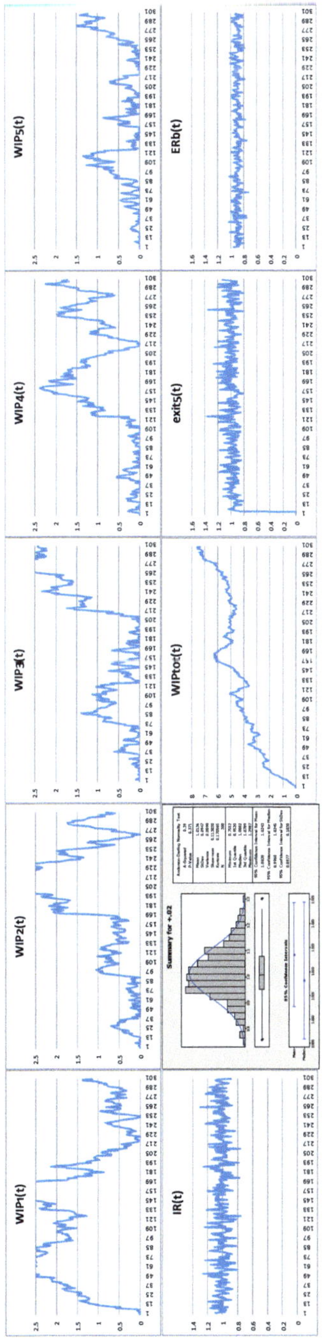

Fig. 4.13 Monte Carlo simulation of WIP formation due to variation in CT and IR with E[IR] = 1.02/min, SD[IR] = 0.1; E[CT] = 1, SD[CT] = 0.1 simulating a (G/G/1) queuing system; please note, the IR is only slightly above of the mean of the ER of the five steps

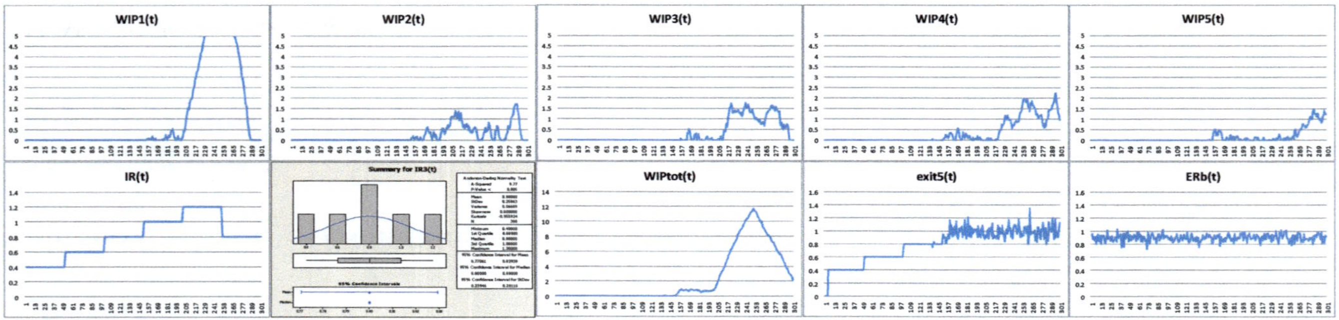

Fig. 4.14 Monte Carlo simulation of WIP formation due to variation in CT and deterministic stepwise change of IR with IR = 0.4 (t = 1..50); 0.6 (51..100); 0.8 (101..150); 1 (151..200); 1.2 (201..250); 0.8 (251..300)/min, E[CT] = 1, SD[CT] = 0.1

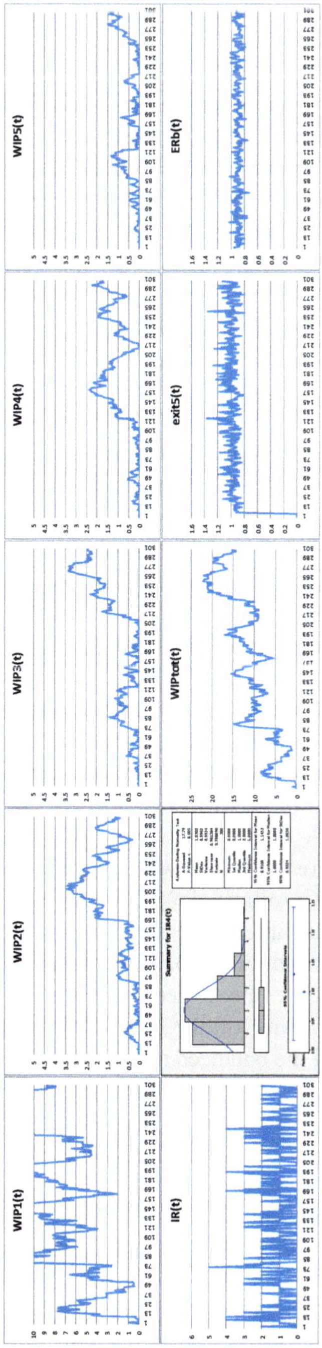

Fig. 4.15 Monte Carlo simulation of WIP formation due to variation in CT and IR with E[IR] = 1/min but with Poissonian (λ = 1) release also to the shopfloor and E[CT] = 1, SD[CT] = 0.1 simulating a (M/G/1) queuing system

non-perfect balanced production line with normally distributed cycle times N(1; 0.1), being in fact a stochastically balanced manufacturing line as usually is the case in reality, and therefore also ER, at each station creating a variable bottleneck along the line. We have therefore not a time-invariant situation as enounced by the second Corollary of Throughput. The output of the line is shown in the chart $exit_5(t)$ and the capacity of the varying bottleneck in the chart $ER_b(t)$.

In Fig. 4.11 the IR is deterministic with 1 liter/t and the average bottleneck of the line is not 1 but $E[1/CT_b = ER_b] = 0.9$. The exits at the end of the line (exit5), after 5 time periods when the water arrives at the end of the line, begins with $E[ER_b] = 0.86$, i.e. with approximate the mean bottleneck ER and after the transitory has expired arriving at a steady state, the exit rate of the line corresponds maximum to the $E[ER] = 1$ of the last station 5 with its specific variability. From Fig. 4.11 we see that the WIP increases for the first 100 cycles and stabilizes from cycle 180 onwards between 3 and 3.5 liter. The stabilization of the WIP is a direct consequence of the Corollary of Weak WIP Stationarity only after transitory has expired (please note, the weak stationarity is not perfect because the CT are not deterministic but show variation) but this is due to the fact, that the formation of the WIP in front of each operation tends to decouple the operations. This decoupling allows then each station to work at its full, but variable, capacity limit equalizing the IR in the average, leading then to the applicability of the Corollary of Weak Stationarity. Indeed, after transitory, i.e. after approximate 50–100 cycles the whole WIP of the line tends to stabilize at 3–4 liter (or pieces, if you want). During the transitory period (1–100), the average exit5 is 0.92 and at steady state (101–300) the average exit5 is 1.0, i.e. at the nominal average capacity rate. This is the reason why automotive TFL usually have a small FIFO buffer in front of each operation in order that issues at a workstation does not repercute downstream with the consequence of limiting throughput. WIP zero in between operation is usually only implemented for high performance automatic high speed takted TFL. It is also interesting to see, how WIP is spreading evenly over time between the five operations as can be seen by Fig. 4.16 according to the Corollary of WIP Spreading.

Figure 4.12 shows still the same process but now also the IR has stochastic characteristic with normal distribution equal to the mean and variation of the cycle time of operations. We observe the expected behavior, with increasing exits attaining after transitory the exits of the last section. Interesting, such as in Fig. 4.11, i.e. the IR equalizing the balanced capacity, but now not deterministically but with stochastic variability, the WIP stabilizes in a more erratic way, at a slight lower level of 3 than before; the reason is that the average of the input rate of 300 cycles is due to the output of the random generator giving $E[IR] = 0.995$ however before it was $IR = 1$ exactly.

Figure 4.13 has the same process with the IR having again a $SD[IR] = 0.1$ but now the $E[IR] = 1.02$, i.e. very little above of the average capacity of the line, and in fact, the WIP, although not much, but it is increasing steadily, the process not being able to digest the incoming quantity according to Eq. (3.7c).

Figure 4.14 has once more the same process but with a step varying IR beginning with 0.4 liter/t, i.e. much below of the nominal capacity of 1 liter/t and increases by

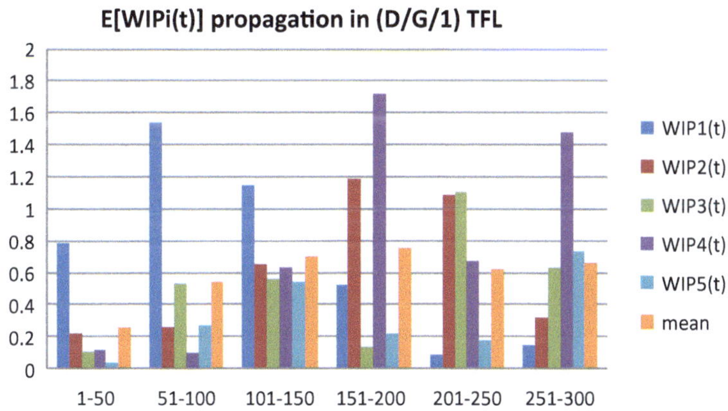

Fig. 4.16 Spreading of WIP across the operations

0.2 liter after each 50 cycles. Interesting is, that until the average bottleneck capacity of 0.9 liter is not reached, no WIP is build-up and the exit at the end of the pipe corresponds to the input rate. Reaching the IR = 0.8 (from t = 101..150) versus the $E[ER_b] = 0.9$ it becomes not yet critical but might experience the variation of the CT and from t = 151..200 with IR = 1, i.e. larger than 0.9 we observe forming of WIP. The WIP grows faster when the IR climbs to 1.2 liter during t = 201..250 and melts only when the IR goes back to 0.8 liter (t = 251..300). Also in this case, as soon as the WIP forms and the pipe sections are decoupled, the exits at the end corresponds to the exit of the last section with $E[ER_5] = 1$ and its variability.

In Fig. 4.15 we still have the same process with normally distributed cycle times, i.e. exit rates, but now we have a Poissonian-type release of production orders to the shopfloor. This release pattern is quite typical in traditional B&Q manufacturing realities, i.e. according to order load. The effect is clearly visible in the simulation. The WIP is forming mainly in front of the first operation congesting the production area. Although the average of the release corresponds to the nominal capacity, the high variance is reflected also in the total WIP. The alternating evolution is due to the characteristic of Poissonian distribution that comprises also no production orders that are released in certain instants.

These simulations show how sensitive the process behaves when IR is approaching the nominal production capacity and how variation in cycle time affects the output. From there derives the attention to reduce cycle time variability, hence, what the Japanese call Mura.

Let us now simulate a new process with the same $E[CT_i] = E[CT_{i+1}] = 1$ but with higher variability such as a Markovian exponential distributions. This situation might represent very well fully flexible Industry 4.0 manufacturing systems, full flexibility being one of the objectives [6, 7]. Please note, to limit extreme peaks of the right skewed distribution tail, outliers have been excluded.

In the last two simulations of Figs. 4.17 and 4.18 we try to simulate the behavior of fully flexible Industry 4.0 manufacturing systems without having redundant capacities and lack of central control unit to be better comparable to the former simulations. Please pay attention, the high variability of CT is also typical of most of today's shopfloor equipment where rather a technological (make-to-order) than a logistical (make-to-stock) production capacity is offered to customers.

In Fig. 4.17 we show the simulation of $E[CT_i]$ stochastic balanced manufacturing line but with high variance ($1/\lambda^2$) following an exponential distribution. The stepwise-change of discrete IR allows to evaluating the behavior in different conditions. With $IR(t = 1..50) = 0.4$ and $IR(t = 51..100) = 0.6$, the system does not have problems to process the arriving orders although the $E[ER_b] = 0.2$, because the forming WIP is decoupling the process steps. The formation of WIP begins already with an $IR = 0.8$, i.e. below nominal capacity. Indeed, the E[exit5] is stable at an average value of 0.98 exits per cycle. As soon as the nominal capacity is reached, i.e. reaching $IR = 1$, WIP growths and explodes; in reality this would not happen, because arising the need to debottleneck the system with additional capacity.

Figure 4.18 reflects per antonomasia the hypothetic performance of not yet optimized fully flexible Industry 4.0 manufacturing system. Non-deterministic Poissonian IR of highly erratic orders, direct consequence of IOT, eventually with bulk arrivals and immediately released to shopfloor. The high variance of CT reflects the technological capacity of the workstations to perform "ad hoc" customized manufacturing of different content. This structure will be typical of fully flexible Industry 4.0 systems with addressable independent workstations by AGV [6, 7]. Figure 4.18 shows that the WIP explodes although the Corollary of Weak WIP Stationarity applies. This shows that the corollary is very sensitive to the equal sign and does not tolerate too excessive variation. This is for sure a handicap of such systems. What can be done? The situation can be controlled by additional workstations, i.e. investing in additional capacity. Concomitant arriving orders may be optimized by scheduling sequence. But all this needs a central ERP 4.0 CPPS (cyber physical production system) control unit. Indeed, this situation corresponds to a multi-objective optimization problem where Pareto optimality is the target in order to match capacity as well OTD to the customer. Pareto efficiency, i.e. solutions belonging to the Pareto frontier, will be the measure to minimize Muda by balancing the requirements of the SPQR axioms. Envisaged neuronal decentralized intelligence will not be sufficient to optimize the whole; sophisticated optimization algorithm have to be developed as a central control unit to coordinate decentralized intelligent units to make such a flexible production system viable. Generally, high variation most probably has to be limited also in Industry 4.0 manufacturing systems.

The insights given by these simulations show mainly two facts: the deleterious effects of CT variability and the beneficial side effect of WIP (of course if stable and contained) to allow to exploit the capacity limits of all process steps. The latter one is direct proportional to PLT and inverse proportional to OTD fulfillment, if no capacity adjustments are implemented. The experienced performance was for a

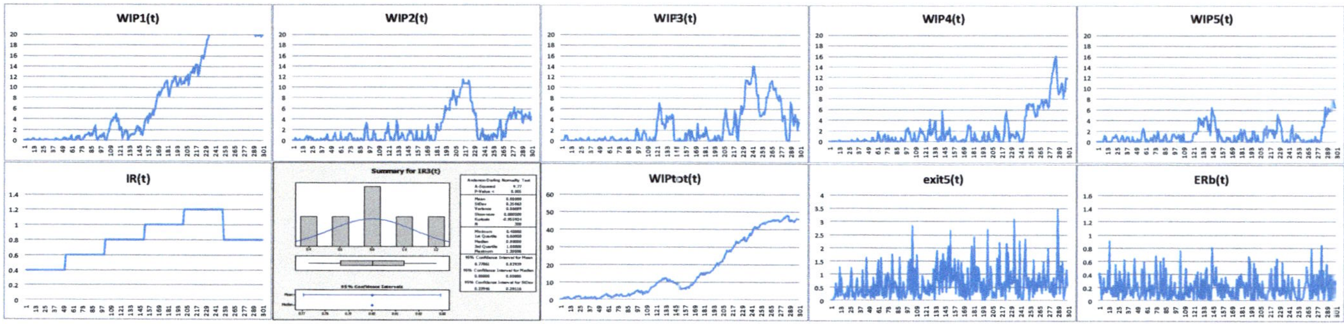

Fig. 4.17 Monte Carlo simulation of WIP formation due to CT and IR variation. IR with discrete stepwise variation. NB: E[ER] = E[1/CT] of Markovian distribution-type with $\lambda = 1$ representing a (G/M/1) queuing system

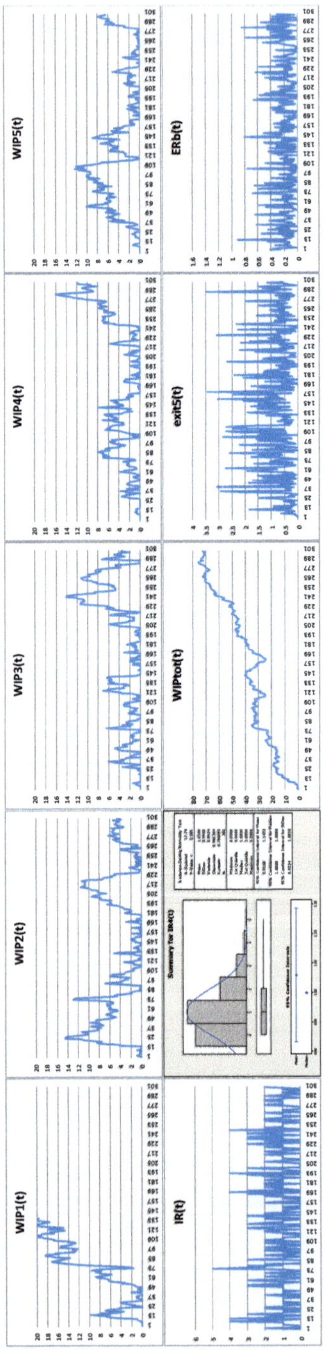

Fig. 4.18 Monte Carlo simulation of WIP formation due to CT and IR variation. IR with discrete Poissonian IR with λ = 1 and CT of Markovian distribution-type with λ = 1, i.e. representing a (M/M/1) queuing system

balanced line with stochastic variation in the cycle time; the situation worsens in the case of an unbalanced line showing also stochastic variation in CT. We can now enounce the

Lemma to the Theorem of Stochastic Cycle Time Variability (Lemma of Minimizing CT Variability or Pull Recommendation)

Given is a stochastically CT-balanced single piece pull (SPpull) manufacturing line. To avoid a drop in ER during the transitory ramp-up (each time a new batch is launched), the variability of the CT_i has to be limited, being the ER inverse proportional to the longest CT (upper natural limit of variability).

Corollary to the Theorem of Stochastic Cycle Time Variability (Corollary of Natural Operation Decoupling or Push Effects)

When orders, i.e. takt rate TR, is approaching the capacity of a stochastically CT-balanced SPF line, the variability of CT will generate WIP between the randomly created time traps and after expiring of the transitory, the WIP will have decoupled the operations allowing to operate each operation at its average capacity.

The Lemma of Minimizing Single CT Variability is a necessary but not sufficient condition to observe the Corollary of Strong WIP Stationarity (white box), because not an attainable condition. Strong WIP stationarity is only attainable in automatic high performance TFL; the weak WIP stationarity (black box) receives therefore a high practical relevance. The question is now, how much variability is sustainable. This question is not only important for the performance but also to seize the buffer in front of a queuing system which does not have infinite buffer capacity b as assumed until now, therefore having a queuing system of (D/G/1/b) with limited buffer capacity. From Eq. (3.7b) with $E[CT_{i-1}] = E[CT_i]$ we can write

$$dWIP_i = \left(\frac{1}{E[CT_{i-1}]} - \frac{1}{E[CT_i]}\right) \cdot dt = 0$$

$$dWIP_i = \left(\frac{1}{SD[CT_{i-1}]} - \frac{1}{SD[CT_i]}\right) \cdot dt \neq 0$$

resulting in an increasing or decreasing WIP according to variation of the cycle times and the potential existence of a non-foreseeable "structural-random" bias in the stochastically distributed cycle times. We will not further simulate here this situation and leave the exercise to the interested reader. We can conclude this section by stating for practitioners with different words:

- the average capacity of the workstations of a manufacturing line is important for the long run stability of the WIP, i.e. reflecting the black box stability
- whereas the variance of the workstations CT is important for the short run stability, reflecting the white box stability.

References and Selected Readings

1. Suzaki, K.: The New Manufacturing Challenge. Free Press, New York (1987)
2. Rüttimann, B.G.: Discourse about linear programming and lean manufacturing: Two different approaches with a similar, converging rational. J. Serv. Sci. Manag. **8**, 85–91 (2015)
3. Rüttimann, B.G., Fischer, U.P., Stöckli, M.T.: Leveraging lean in the office: Lean office needs a novel and differentiated approach. J. Serv. Sci. Manag. **7**, 352–360 (2014)
4. Goldratt, E.: The Goal, Excellence in Manufacturing. McGraw-Hill, New York (1984)
5. Goldratt, E.: Theory of Constraints. North River Press, Croton-on-Hudson (1999)
6. Umsetzungsempfehlungen für das Zukunftsprojekt Industrie 4.0—Abschlussbericht des Arbeitskreises Industrie 4.0. www.plattform-i40.de, April 2013
7. Zukunftsbild Industrie 4.0. www.plattform-i40.de

Chapter 5
Increasing Cell Utilization

In Chap. 4 we have developed the performance metrics of manufacturing and process lead time applied to SPF and B&Q principle enunciating the following theorems for a mono-product cell:

- Theorem of Generalized Lead Time
- Theorem of Lead Time Stability
- Main Theorem of Production Time (SPF dominance)
- Theorem of Debottlenecking
- Central Limit Theorem of Manufacturing CLTM (Identity theorem)
- Theorem of Stochastic Cycle Time Variability.

In the following sections we will develop the concepts of the second level of in-house manufacturing complexity according to the TPS model of Fig. 2.3 dealing with the optimized utilization of manufacturing cells, reasoning now extended to a multi-product manufacturing regime. The main concepts are the workstation characteristics of WTT for a mixed-product model and the batch size, as well as how to design and adapt the capacity extending the concepts to cellular manufacturing.

© Springer International Publishing AG 2018
B.G. Rüttimann, *Lean Compendium*, DOI 10.1007/978-3-319-58601-4_5

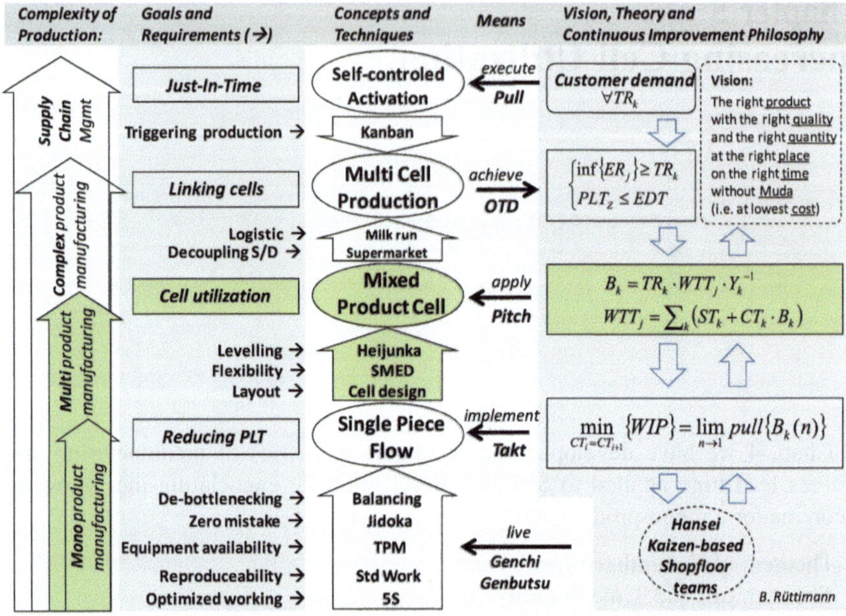

5.1 Product Mix Variability and Heijunka Leveled Scheduling

Before entering into the topic of multi-product manufacturing, we have to think about, how products are allocated to machines and workstations, or grouped into a manufacturing cell, case which is more complex and will be the topic of this chapter. The different ways are described by the **allocation principles**:

- "1-to-1", one product has dedicated one workstation, or even a cell;
- "1-to-m", one product can be manufactured on different similar machines;
- "k-to-1", k different products are manufactured on the same machine or within the same cell;
- "k-to-m", k different products may be manufactured on m different workstations, or even complex cells.

There is a quit different philosophy between Japanese and European allocation mentality. Whereas European production managers show preference for the "1-to-1" principle for cellular manufacturing avoiding setup times (if the production volume is big enough, obviously), Japanese production managers prefer the "k-to-1" principle, as we will discuss. For simplicity reason, we will concentrate on the "k-to-1", to understand the "k-to-m" principle. It seems clear, that if k "1-to-1" cells exist, one for each of the k products, due to parallelization there is no issue to supply the ordered mix punctually, this becomes more complex in the case only one cell with the "k-to-1" principle exist. The latter principle, however, is less capital intensive and allows a better utilization of the equipment. Usually a fabrication plant may have a dedicated line for high volume products (flow shop) as well as a more flexible shopfloor with several workstations for low runners (job shop). We will not enter here into plant conception but limit our analysis for a cell. We will see, how OTD can be implemented despite k concurrent products are manufactured within the same cell (k-to-1).

In a non-customer imposed TR regime, the order rate OR usually follows a queuing distribution of Type (G/D/1) or (M/D/1), i.e. the arrival has stochastic characteristics regarding quantity and product. The distribution of the order arrival rate OR is important but has not to be over-emphasized in importance, because through the formation of backlog and the scheduling operation a random arrival rate is always transformed into a deterministic input rate IR with (D/D/1) distribution, i.e. the release rate into the physical manufacturing of shopfloor level can be optimized. Usually, the incoming orders will be "stacked" into a FIFO organized pile, with a **backlog waiting time (BWT)** of the order intake. Indeed, if we assume a FIFO scheduling principle, the **customer visible time (CVT)** of delivery is given by Eq. (5.1)

$$CVT = BWT + MLT \leq EDT \tag{5.1}$$

i.e. composed of order backlog and order manufacturing time, where MLT is the already modeled manufacturing lead time and EDT is the already seen expected delivery time. To satisfy Eq. (5.1), the Theorem of General Production Requirements (OTD Theorem) applies. The implications of Eq. (5.1) will be further developed in Chap. 7. Apart from rather deterministic expected repetitive product call-offs with TR, the products with stochastic OR, however, have to be produced too and, to the contrary of a deterministic TR with a defined product mix, the order sequence and machine setups have to be scheduled in order to maximize equipment utilization. After having made a quick excursion to traditional scheduling, we will start first in this section with developing for simplicity reason the deterministic customer mix.

To avoid long CVT, Lean theory is not intended to experience the whole production time chain as to Eq. (5.1), Lean is minimizing BWT or even eliminating BWT reducing Eq. (5.1) to

$$MLT \leq EDT \tag{5.2}$$

and uses a Heijunka scheduling optimized sequence for JIT fabrication as we will see. A workstation manufacturing different products is symbolized in VSM by Fig. 5.1.

Different from the operations research approach to optimize e.g. an annual production or weekly scheduling campaign, i.e. defining the optimal mix for maximizing overall profitability, such a workstation usually has to supply a given mix in JIT. Fabricating a mix of k products and not a single product makes it necessary to introduce the **workstation turnover time (WTT)** as well as the

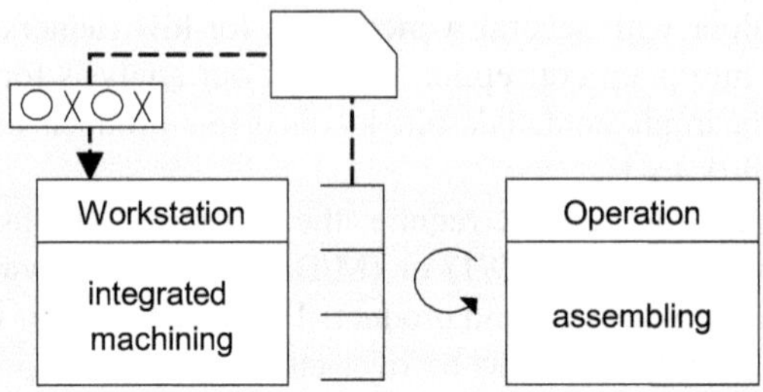

Fig. 5.1 "k-to-1" workstation with Heijunka-box leveled product-mix

product interval time (PIT). The PIT is the interval time until the same product is manufactured again and represents a kind of frequency. This PIT is very important to calculate the Kanban max loop, as we will see.

The $WTT_{i|k}$ is defind for a single workstation i processing k similar, but different products. When the workstation switches from a product k to a product k + 1, a setup of the equipment has to be performed to allow the changeover to the next product. The time needed to perform the setup is called **setup time** $ST_{i|k}$ of the workstation i for the product k. The $WTT_{i|k}$ is given by Eq. (5.3) and is the time to manufacture once all products with the batch size B_k and pertinent cycle times $CT_{i|k}$ including setups

$$WTT_{i|k} = \sum\nolimits_k \left(ST_{i|k} + CT_{i|k} \cdot B_k\right) \tag{5.3}$$

The WTT is independent of the scheduling order of the various products composing the product family processed on the workstation and is a characteristic of the workstation as it is shown in Fig. 5.2 (the planned sequence is usually given by the Heijunka box scheduling).

Further, the timeframe until the same product k is produced again is called product interval time PIT_k given by Eq. (5.4) and represents the inverse of the frequency it is produced. It depends from the batch size B_k and its **yield** Y_k as well as its takt rate

$$PIT_k = \frac{B_k \cdot Y_k}{TR_k} \tag{5.4}$$

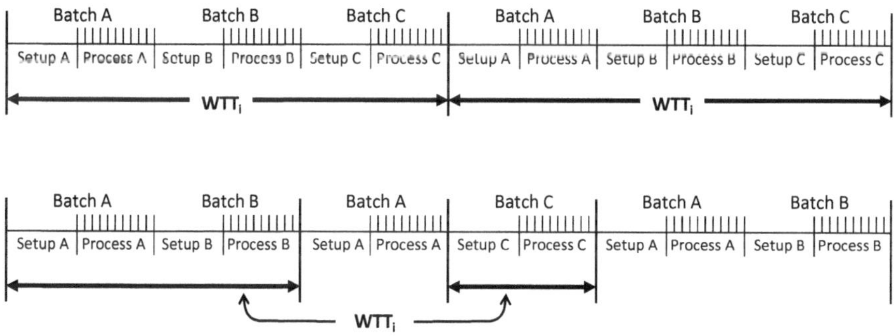

Fig. 5.2 Graphical representation of WTT for different scheduling sequences [based on 1]

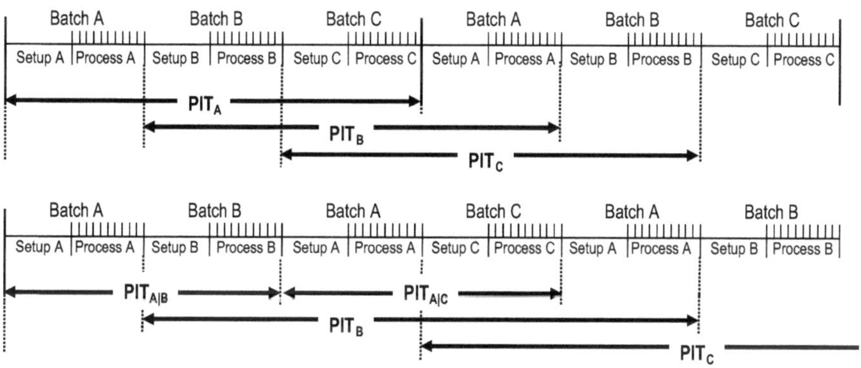

Fig. 5.3 Graphical representation of PIT for different scheduling sequences

The TR is the rate of demand how fast the quantity B_k put to the inventory (supermarket) is consumed and the PIT_k the time span until it is consumed. Instead of developing PIT in function of B_k and the other variables, the correct question should be: how large has to be the batch B_k to bridge the PIT, i.e. the interval time until the product k will be produced again. Equation (5.5) results from Eq. (5.4)

$$B_k = \frac{PIT \cdot TR_k}{Y_k} \tag{5.5}$$

The question about sizing the optimal lot will be developed in the next section. Please note, whereas WTT is an invariant characteristic of the manufacturing cell (when the mix is given the WTT is defined) not depending from the manufacturing sequence, however the PIT depends from the manufacturing sequence and may therefore vary as shown in Fig. 5.3.

Please note, if the pitch is not equal to the standard Heijunka timeslot but to a logic (dummy) timeslot, then $PIT_{A|B}$ is not equal to $PIT_{A|C}$ of alternating scheduling. Indeed, the PIT is a characteristic of the product; further, the WTT is a characteristic of the workstation, and for completeness, the PLT is a characteristic of the process, as well as MLT is, but the later depending also on the manufacturing modus, i.e. the transfer principle. Before extending the concept to a manufacturing cell let us now put the attention to scheduling principles of a workstation.

Let us suppose, the deterministic mix to be delivered each day might have a different requirement with regard to the quantity of each product to be manufactured, i.e. e.g. 200 electric motors for the two front side windows, 100 for the big front wiper and 100 for the right rear mirror, assuming the motors are not identical necessitating changeovers, i.e. new setup of the workstation. This unevenness in the demand from 500 to 1000 pieces in the case of weekly deliveries, the Japanese call it Mura, creates Muri and both are in turn a major source of Muda. Generally, in a manufacturing environment the utilization of production equipment has to be exploited to the maximum and therefore the machines have to process several products with different CT_{mlk} as it is usually encountered in reality. This implies scheduling of sequence and machine setup to allow the changeover to another product. In western high performance thinking paradigm, the scheduling is usually made for weekly production campaigns. The batches are as large as possible to minimize the downtime of machines during setup with high WIP but also generating high finished goods inventory (FGI), being only possible to supply the mix next week when all products are jointly available. In addition, this causes long batch MLT and low flexibility, and other products needing to wait increasing BWT until it is their turn to be produced in a FIFO scheduling (Fig. 5.4).

The average FGI (independent which type of motor) with weekly delivery of aggregated 2000 pieces is 1000 pieces (2000/2). Toyota followed a different approach. Briefly, unevenness (Mura) introduces additional overburden of information to be controlled (Muri) which finally causes inefficiencies (Muda); for further explanations we refer to e.g. [2–4]. Not push-scheduled manufacturing

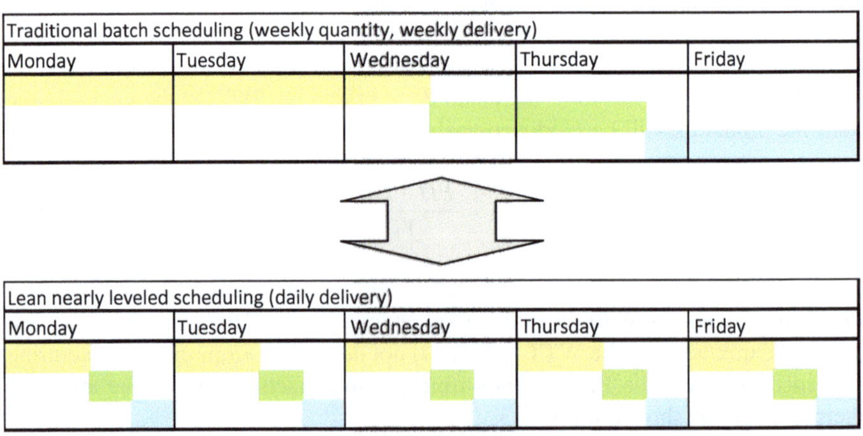

Fig. 5.4 Traditional non-leveled weekly production scheduling versus daily scheduling for three products (please note, the second scheme is not yet pitch leveled but follows the Lean mantra "make every product once a day")

principle, but consumption-oriented customer pull is the solution to reduce FGI. In addition, to increase flexibility (here intended for a given product mix to be produced) quick setups are important to change over from one product to another. To speed-up MLT, the batches have to be reduced diminishing again FGI. This leads to a synergic system with lowest WIP, fastest lead time and JIT supply to each workstation with finally respect of OTD. The aim is to produce at least every product once a day (Fig. 5.4). The result of daily deliveries of 400 pieces motor mix (to assemble a car one needs a set of motors) is a reduced FGI of average 200 pieces, requiring also less physical storage space and holding cost. To level also the intra-day production, a pitch is defined. Pitch is the smallest defined standardized manufacturing batch, which by the way is a multiple of the pack-out quantity, matching usually the logistic transportation unit. This allows to responding quickly to changing demand observing the highest flexibility. Moreover, the aim is to manufacture several times a day the product by the aid of a Heijunka-box. A Heijunka-box divides the time of the day in equal time slots representing the time pitches (Fig. 5.5). The scheduling is therefore not deterministically planned, but an instantanious reaction triggered by the random demand, i.e. matching the downstream call-offs. The cell becomes a self-controlled planned unit.

The pitch, i.e. the Heijunka timeslot for the example of Fig. 5.5, is calculated for the workstation X with three products A, B, and C having each, e.g. CT = 66 s with each having a pack-out quantity (handling and shipping unit) of 50 pieces, which corresponds to the batch $B_k(1)$ and a workstation changeover time of 5 min

$$timepitch = ST + CT \cdot B_k = 5\,\mathrm{min} + 66\ \mathrm{sec}\ \cdot 50 = 60\,\mathrm{min}$$

i.e. a time-pitch of 60 min. The WTT corresponds therefore in this case to 180 min. This allows with alternating scheduling and double frequency of A to have two shipments within one shift with the mix A, B, C. This replenishment corresponds to the fixed-quantity fixed-interval principle, as we will see in Chap. 7.

Please note, that the PIT_B=PIT_C, but the PIT_A is half of PIT_B and PIT_C allowing to have the same lot quantity leading to a leveled fabrication load.

Heijunka Box of Cell X/ABC										
Part	Daily production requirement	06:00	07:00	08:00	09:00	10:00	11:00	12:00	13:00	14:00
A	200	50		50		50		50		
B	100		50				50			
C	100				50				50	
additional Kanban										

Fig. 5.5 Heijunka-box pitch-leveled production of cell X allowing two shipments in one shift (pitch quantity is 50 pieces)

Lemma of Scheduling a Mixed-Product Workstation or Cell (or Lemma of Heijunka)
The leveling of the production mix is done with the help of a Heijunka-box by defining a pitch, i.e. identical time slots, in order that the workstation's (or cell's) products can be produced alternatively also several times a day.

Of course, these concepts suits best a deterministic product mix. What can be done by a non-deterministic mix? This might look impossible, but by defining dummy time slots (rather spare capacity than time slots) it might become possible (with a little bit of juggling) to implement this also for general production with limited variation.

We can now enunciate the enlarged definition of bottleneck generalizing the Theorem of Throughput for a single product, extended to a mix of products to be delivered jointly JIT.

Theorem of Generalized Throughput (or WTT-Aggregated Bottleneck Theorem)
In a multi-product manufacturing cell (or transfer line) with cycle times $CT_{i|k}$ the workstation (or cell) with the longest WTT_i (or CTT) determines the bottleneck workstation (or bottleneck cell) to supply a mix of products. Please note, the exit rate ER of the cell is still given by the workstation with longest $CT_{i|k}$ processing a certain product.

First Corollary to the Theorem of Generalized Throughput (Corollary of Generalized Bottleneck Time-Variance)
Different than in mono-product cells (or line) where the bottleneck is time-invariant (under "ceteris paribus" conditions), in mixed-product manufacturing cells with high differences between the cycle times $CT_{i|k}$ of different products k, the operation i for a product k being the bottleneck may vary and the bottleneck being therefore time-variant according to the product k in the cell.

Second Corollary to the Theorem of Generalized Throughput (Corollary of Frequency)
The product interval time PIT, i.e. the frequency, until when the same product within a mixed-product workstation or cell is manufactured again, depends from the number of products in the cell and from the pitch as well as the takt rate of the product consumption.

Notice, the PIT, i.e. the frequency depends from batch size B_k and TR_k but not directly from WTT, however B_k depends from WTT and therefore reflecting PIT, as we will see. The WTT is a wider concept of bottleneck and approaches the related concept of bottleneck in Linear Programming of Operations Research [5] corresponding to the resource **b** rather than the longest CT in the matrix **A** of equation system (3.3).

Let us now deal with the concept of manufacturing cell. A manufacturing cell is a self-contained (i.e. product and machine-specific) as well as self-controlled (i.e. operator managed and consumption triggered) production unit; indeed, different from Western technology dedicated "competence centers" (functional fabrication layout which, indeed, shows the incompetence of managers) a cell comprises all necessary machines and workstations to produce a defined family of components. A cell transforms raw material or an intermediate component into a finished product or component with higher added value. Cellular production is triggered by the downstream supermarket-pull to be replenished. Cell organization is following usually a layout allowing a SPF, but not necessarily; in that case, although the work stations are placed near together, the B&Q principle represents a fake flow. In VSM, the cell is an aggregated representation of a sequence of operations as shown in Fig. 5.6. The machines, or equipment, are usually displayed in a U or Z-shaped form in order to favor the moving of the operators within the cell (Fig. 5.7).

A cell performs best if the cycle times of the single operations within the cell are balanced (Corollary of Lead Time Limit). To increase utilization of the cell, several similar products forming a product family are grouped and assigned to the same cell. The necessary and sufficient conditions for a product family will be enounced in Sect. 5.3. In the case of only one product being manufactured in a (dedicated) cell, we call it a mono-product cell (1-to-1) otherwise a mixed (or multi)-product cell (k-to-1) according to the assignment principle. A mixed-product manufacturing cell (allocation principle "k-to-1") can therefore be described by a matrix **A** such as given by Eq. (3.2).

Definition of a Mixed-Product Manufacturing Cell

A manufacturing cell is a self-contained and self-controlled mixed-product manufacturing unit, and therefore separated by a Kanban managed inventory (supermarket) to the downstream operation, for which all precedent theorems, corollaries and lemmas apply; i.e. a process "en miniature", processing different but similar components or sub-parts for the next downstream step.

If several workstations are combined into a cell, things get more complicate than modeled with Eq. (5.3). Such as an entire plant, also each upstream cell of a downstream transfer line has a PLT, a cell PLT. Instead of using the acronym PLT for the cell lead time we prefer to use here the diction **cell lead time (CLT)** which corresponds to the $PLT_{j|k}$ of the cell j for the product k; we prefer to reserve the expression PLT for the entire process, being the cell an aggregated operation part of the main process or confluent process. We will develop the laws supposing to have a SPF with CT being nearly balanced or the operations are balanced through the operators, cell bottleneck are in this case not given by the operation, but by the operator. This assures that the best performance can be expected (Corollary of Lead Time Limit). The best performance will be defined by Eq. (5.6) which corresponds to the **work content (WC)**; notice, WC underestimates the real $CLT_{j|k}$ of the cell j for product k as soon as a non takted WIP is generated

$$CLT_{j|k}^{SPF} \approx \sum\nolimits_{i} CT_{i,j|k} = WC_{j|k} \tag{5.6}$$

In the case the cell j is composed of several work stations i (as it is usually the case) the corresponding turnover time to WTT, which we will name the cell turnover time CTT, the Eq. (5.3) for the workstations becomes for the cell

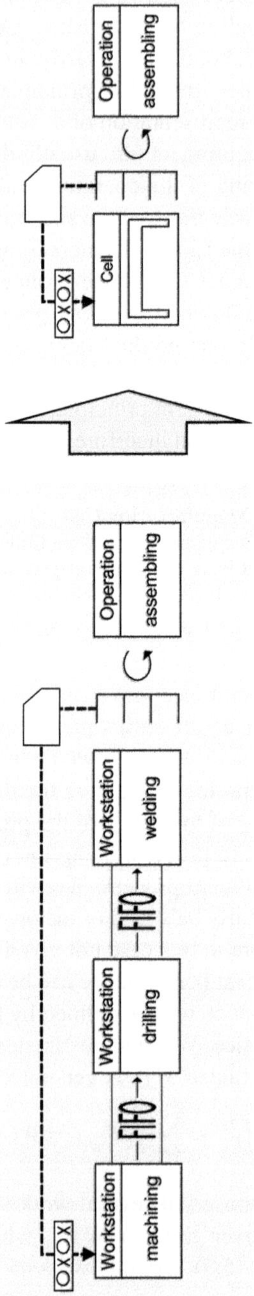

Fig. 5.6 VSM-aggregated representation of a three workstation manufacturing cell with Heijunka-leveled product-mix

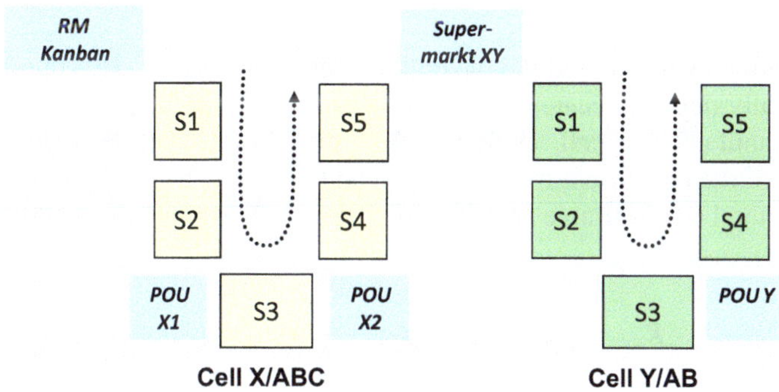

Fig. 5.7 Possible adjacent layout of two consecutive manufacturing cells X and Y linked by a supermarket XY

Eq. (5.7), assuming that the setups are done sequentially when the line is completely empty

$$CTT_{j|k} = \sum_{k} \left(\sum_{i} ST_{i|k} + MLT_{nPF,k} \right) \tag{5.7}$$

The setup times of the various workstations i of the cell j may differ, however we will assume they are equal, which is quite reasonable if the products are similar, otherwise taking the longest setup time if all setups are performed in parallel by the operators in the cell. The products suits best to the pitch if $MLT_{nPF,k}$ is equal for all products k of the cell.

Please note, different than sometimes superficially treated and discourse limited to WTT in literature, the WTT depends not only from the workstation but also from the transfer principle if the workstation is not a single machine but a cell composed of several steps, and therefore MLT_{nPF} (4.9a, 4.9b) and not CT applies, as we will see. The more complex the cell is, the more important becomes Eq. (5.7). Indeed, the distinction between single workstation and cell (sequence of multiple workstations) for the lead time calculation is necessary, because on a single workstation $MLT_{SPF} = MLT_{B\&Q}$ is confounding whereas in a cell

$$MLT_{SPF} \neq MLT_{B\&Q}$$

The corresponding PIT for the products of the cell is identically to that of the workstations, because the PIT depends from the product batch and from the consumption rate, which is usually the TR. The calculation of the PIT for the cell is therefore identically to Eq. (5.4) for the workstation. What is different is the CTT as we have seen. The question now arises how to calculate the optimal batch size to bridge PIT.

5.2 Lean Batch Sizing

The question now is: how should the optimal batch been sized? This question is not unequivocally defined because the question arises only in the presence of a multi-product manufacturing cell, or line, and relative to fix cost, i.e. e.g. the related downtime of the equipment for the setup. The traditional batch sizing relies on the concept of total cost, variable cost and holding cost. The total cost function is given by Eq. (5.8)

$$K_{tot} = k_{\text{var}} \cdot Q + \frac{Q}{B} \cdot k_{fix} + \frac{1}{2} \cdot k_{\text{var}} \cdot B \cdot i_{WACC} \tag{5.8}$$

where i_{WACC} represents the weighted average cost of capital, or other holding cost, and the ratio Q/B is the number of times the setup is performed to produce the quantity Q. Applying the first order condition to calculate $\min\{K_{tot}\}$ obtaining finally Eq. (5.9) of economic optimized batch quantity

$$\min\{K_{tot}\} = \frac{dK_{tot}}{dB} = -\frac{Q \cdot k_{fix}}{B^2} + \frac{k_{\text{var}} \cdot i_{WACC}}{2} = 0$$

$$B_{opt} = \sqrt{\frac{2 \cdot Q \cdot k_{fix}}{k_{\text{var}} \cdot i_{WACC}}} \tag{5.9}$$

Equation (5.9), however, does not take into consideration that also other products have to be manufactured at the same time and does not reflect real requirements in a Lean environment. Today, Eq. (5.9) can be considered to be obsolete. Indeed, one of the goal of Lean is to minimize inventories, allowing inventories only when strictly necessary due to the cardinal goal of JIT manufacturing (called then strategic buffers). The existence reason of Eq. (5.9) is originated by k_{fix}; eliminating k_{fix} the batch size may tend to one according to Eq. (5.10)

$$\lim_{k_{fix} \to 0^+} B_{opt} = \inf\{B_{lean}\} \approx 1 \tag{5.10}$$

This is the reason, although applying an ex-post rationalization, why quick changeover such as single-minute-exchange-of-die SMED techniques had been developed by Toyota, or more precise, by the father of this concept, Shigeo Shingo.

Let us now orient towards Lean batch-sizing for practical realities. We will distinguish again the simple workstation from the manufacturing cell. This is necessary, because the workstation as single operation is easier to understand and to perform; the manufacturing cell has not a cycle time but a cell lead time CLT, which is the sum of different cycle times at the workstation composing the cell plus eventually a WIP with queuing time. For the workstation we have to combine Eqs. (5.3) and (5.4).

The calculation of B_k is simple for the case of an equally alternating manufacturing sequence being workstation turnover time and workstation's product interval timc equal as shown in Eq. (5.11) and a little bit more complicate for the unequal case of Eq. (5.12). Let us calculate B_k for the case with equal PIT_k and WTT of Eq. (5.11)

$$WTT_i = PIT_A = PIT_B = PIT_C \tag{5.11}$$

$$WTT_i \neq PIT_k \neq PIT_{k+1} \tag{5.12}$$

In the case of Eq. (5.11) we first substitute PIT of Eq. (5.4) with WTT and transform to obtain B_k

$$B_k = \frac{WTT_i \cdot TR_k}{Y_k} \tag{5.13}$$

Equation (5.13) returns the batch size for product k and contains now the unknown WTT which can be calculated by substituting B_k of Eq. (5.3) with the expression in Eq. (5.13)

$$WTT_i = \sum_k ST_k + \sum_k CT_{i|k} \cdot \frac{WTT_i \cdot TR_k}{Y_k}$$

obtaining

$$WTT_i = \frac{\sum_k ST_k}{1 - \sum_k \frac{CT_{i|k} \cdot TR_k}{Y_k}} \tag{5.14}$$

Equation (5.14) contains now all known variables to calculate the batch size. Please note, in order to observe the non-negativity requirement of the denominator in Eq. (5.14) it has to result

$$E[CT_k] < \frac{TT_k \cdot Y_k}{k} \tag{5.15}$$

Equation (5.14) is therefore only the necessary but not the sufficient condition for the WTT. The necessary and sufficient condition to manufacture several products within the same workstation requires reduced cycle times to comply to takt as shown in equation system (5.16).

$$\begin{cases} WTT_i = \sum_k ST_k \cdot \left(1 - \sum_k \frac{CT_{i|k} \cdot TR_k}{Y_k}\right)^{-1} \\ E[CT_k] < \frac{TT_k \cdot Y_k}{k} \end{cases} \tag{5.16a, 5.16b}$$

Equation (5.16a) can now be put into Eq. (5.13) to calculate the batch size. In the case of Eq. (5.12) with unequal PIT_k a normalized PIT equalizing WTT can be defined, e.g. for Fig. 5.3 by setting

$$PIT_{A|B} = PIT_{A|C} = PIT_A$$

and $PIT_B = PIT_C$

getting e.g. for product A $WTT_i = k_A \cdot PIT_A$

Usually, the batch size calculated with Eq. (5.13) is smaller than calculated with the traditional approach of Eq. (5.9).

Things are getting more complicate in the case of a manufacturing cell. In the case that

$$CTT_j = PIT_k for \forall k$$

then

$$B_k = \frac{CTT_{j|k} \cdot TR_k}{Y_k} \tag{5.17}$$

where $CTT_{j|k}$ for a CT balanced cell j with i workstations for a general nPF is according to Eq. (5.7)

$$CTT_{j|k} = \sum_k \left(\sum_i ST_{i|k} + (B_k(n) - n + n \cdot i) \cdot CT_k \right)$$

$$CTT_{j|k} = \sum_k \sum_i ST_{i|k} + \sum_k \left(\frac{CTT_{j|k} \cdot TR_k}{Y_k} - n + n \cdot i \right) \cdot CT_k$$

$$CTT_{j|k} = \sum_k \sum_i ST_{i|k} + CTT_{j|k} \cdot \sum_k \left(\frac{CT_k \cdot TR_k}{Y_k} \right) + \sum_k (-n + n \cdot i) \cdot CT_k$$

and after some algebraic transformations the cell turnover time for the cell j producing the k products with an nPF transfer principle becomes

$$CTT_{j|k} = \frac{\sum_k \sum_i ST_{i|k} + n \cdot (i-1) \sum_k CT_k}{1 - \sum_k \frac{CT_k \cdot TR_k}{Y_k}} \tag{5.18}$$

Equation (5.18) shows, that regarding the denominator of the cell it is similar to that of the workstation of Eq. (5.14) but the condition (5.15) must be different. Indeed, condition (5.15) becomes Eq. (5.19)

$$E\left[CT_{b|k}\right] < \frac{TT_k \cdot Y_k}{k} \tag{5.19}$$

i.e. the cycle time CT_b of the bottleneck workstation of the cell has to be faster by the factor k compared to the takt time, this in the general case if the cell is not a workstation CT balanced cell. In addition, the CTT comprises also the time frame that the pitch quantity (approximate to the batch size) needs to go through the i workstations of the cell. The result of Eq. (5.18) can now be put into Eq. (5.17) to calculate the batch size of product k which in the case of equal TR have the same lot size.

We can now enounce the

Theorem of Lean Batch Sizing (or Optimal Batch Theorem)

The optimal, i.e. minimum, batch size of a product manufactured within a mixed-product workstation (or manufacturing cell) depends on the workstation turnover time WTT (CTT) and the takt rate TR.

Corollary to the Theorem of Lean Batch Sizing (Corollary of Cycle Time of a Mixed-Product Workstation)

Necessary condition that the replenishment of the supermarket from a mixed-product workstation (cell) assures the availability of all products, is that the cycle time CT of the workstation (CT of the bottleneck within a cell) corresponds to the fraction of takt time TT, where the fraction is aligned with the number of products being manufacturing within the workstation (cell).

5.3 Cell Design

Cell design is different from "large line" or transfer line design. The conception of a line applies rather for automated fabrication of large parts assembly where every workplace usually is staffed by operators or robots and is usually deterministically paced. Further, we will not talk here about flexible manufacturing systems (FMS) of automated cells. These are the logic consequences of former CAD-CAM integration leading to CIM, although the enunciated theorems also apply. For FMS we refer e.g. to [6]. A manufacturing cell is a U-shaped line, usually operator staffed, and rather apt for manual manipulation of rather small parts, although not exclusively. Their common denominator is the concepts of

- process sequence
- takt time
- and WIP.

Cell design is not complex but it necessitates to observing certain doings. A cell is characterized primarily by four aspects which will be takled under a–d):

(a) Define product families:

- define homogeneous product families for the cell
- define logic sequence of manufacturing operations

The products suiting into the same manufacturing cell are intuitively given: they have to go through the same operations. However this is only the necessary condition (although a skipped operation may still be possible and the product does not have to be excluded to have it manufactured in the same cell). However, in a high performance cell the necessary and sufficient condition is, that the single operations have to be scalable. Translated to mathematics this writes for the k products like

$$\alpha_1 \cdot CT_{\bullet 1} + \alpha_2 \cdot CT_{\bullet 2} + \ldots + \alpha_k \cdot CT_{\bullet k} = 0 \tag{5.20}$$

where $CT_{\bullet k} = \overrightarrow{CT_k}$
are the products k represented as vectors. The dot stands for the whole line because the writing [CT_{mk}] might confuse to be a matrix for non "professionals", but the number of machines m of the cell defines the dimension of the vector space. Further, in Eq. (5.20) α is a scalar value and, as we have seen, the $CT_{\cdot k}$ represents a product k of a cell, i.e. the column vector of the matrix **A** of Eq. (3.2). Equation (5.20) reads like, the vectors of a set are said to be linearly dependent, if one vector of the set can be expressed as a linear combination of the other vectors. This leads us directly to enounce the

Theorem of Cell Product Congruency (or Linear Dependency Theorem)
Necessary condition to put similar but different products undergoing the same treatments/operations into the same manufacturing cell is that the cycle times CT_{mk} elements of the cell matrix have to be linearly dependent.

Corollary to the Theorem of Cell Product Congruency (or Corollary of "Imperfect" Dependency)
If a product of a family of products with linear dependent CT is of smaller dimension, i.e. it has all elements (operations) of the vector except one, it may remain within the cell, skipping the missing position (operation).

(b) Understand work content:

- define work content of each operation and measure cycle time of each single operation
- compare autonomous machine cycle time and operator work cycle
- automate wherever possible (reach minimum level 3 of automation [7])
- identify bottleneck and level it out if possible

To attain level 3 automation implies to have no Muda in the manufacturing cell regarding waiting times of machine and operator, machine and operator being synchronized.

(c) Conceive layout of the cell for flow:

- layout the cell for flow (avoid fake flow) and observe EHS aspects
- define the WIP and its location

This step is a direct consequence of the application of the Lemma of SPF Regime. The cell can be designed according the following **layout principles**: U-cell, Z-cell, L-cell, I-cell; which one will be applied is contingent to space, obstacles, and staffing. Notice, for cell design this might not generate too much trouble to arrange equipment and workstations to allow a SPF. However, if we extend the concept of cell design to large dimension of existing layout organized in techno-functional aggregated workshops to be linked, no flow will be possible if not a new greenfield operation is planned. For brownfield operations neither operational research algorithms of transportation type nor graph optimization (apart of transportation waste) realistically apply due to the constraint of technologically requested processing sequence. To break out of this deadlock, only new lean conceived greenfield production sites may guarantee a survival in today's globally increased competitive environment.

(d) Organize the cell:

- define organization and balancing principle (fixed zone, dynamic zone, circuit, also called round robin or carrousel)
- staff the cell
- balance work content between operators for non-circuit organization

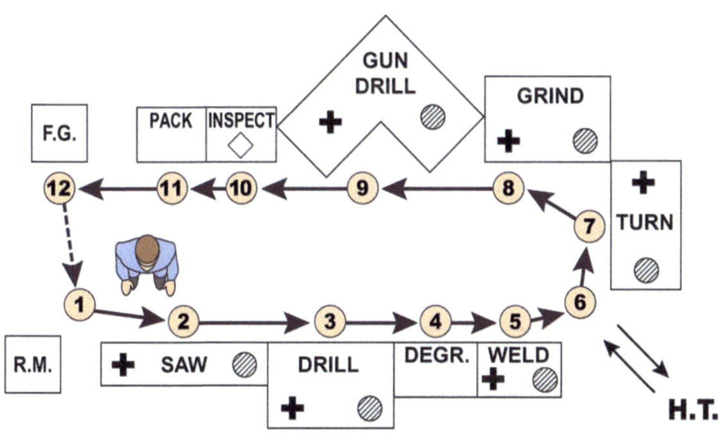

Fig. 5.8 Cell cycle times of workstations and layout is at first conceived for single person staffing [from 8]

A manufacturing cell is always conceived to work with a single person staffing first (see Fig. 5.8) and subsequently organized and staffed with the necessary number of workers.

The CT of the single steps to be performed within the cell has ideally to be balanced; if the balancing of the operation is technically not feasible, the operator's work is balanced. We distinguish following main **balancing principles**:

- "fixed zone", each operator has its (not necessarily consecutive) task to accomplish; this is ideally for a U-shaped cell
- "dynamic zone", each operator has its zone assigned but can invade the up-or downstream zone, if the neighbor operator's pace is not anymore takt aligned; this is ideal for an I-line or L-configuration
- "circuit", each operator is moving along the cell executing each operation of the full work content WC; this is ideal to increase ER of the cell in the case of variable TR but a cell usually cannot host more than three to four operators
- "static", each operator has its place and executes the full WC, easy to increase capacity if WC is operator dominant; this is typical in service operations such as bank teller or supermarket cashier (see Fig. 5.9)

We will briefly enter more into detail for the last mentioned topic (d) of organizing the cell. The staffing of the cell, i.e. the number of workers needed to achieve a certain takt rate TR, can be calculated with the help of the following steps:

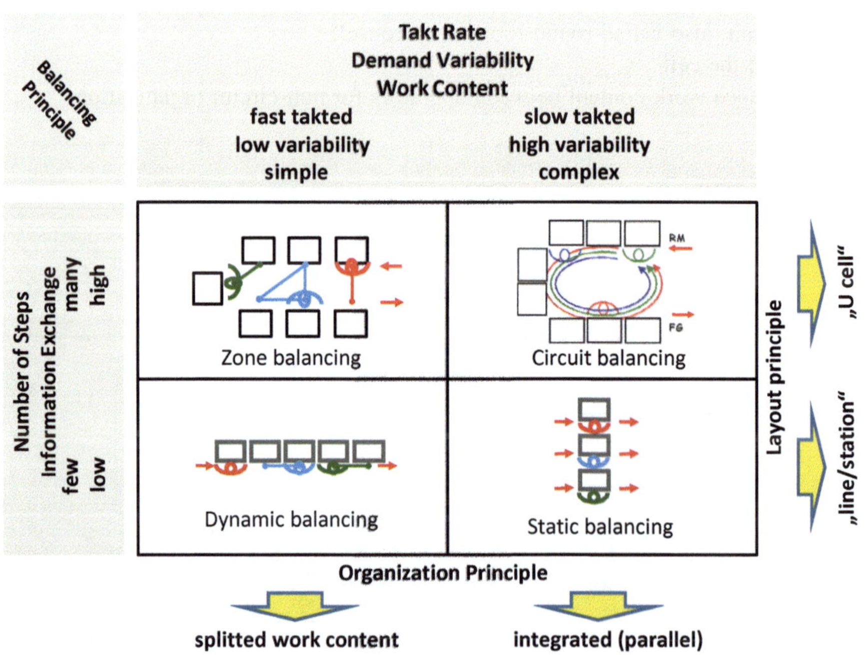

Fig. 5.9 Cell organization and layout concept leading to different balancing principles

(d1) Compute average takt time TT for mixed-product cell, where T_{avail} is the net available time, e.g. in one shift per day minus the pause and setup time divided by the quantity q of the product mix to be produced. The TT may also be imposed by the customer's TR for OTD with JIT characteristics which then requires to adapt T_{avail}

$$TT_{avg} = \frac{T_{avail}}{\sum_k q_k}$$

(d2) Compute average cycle time CT_{avg} where the cycle time CT_{avg} has to be smaller than TT_{avg} by calculating the total time T_{tot} needed per each product k first. Notice, that if the CTs are not perfectly balanced, the $CT_{m|k}$ should be that of the bottleneck of the cell

$$T_{tot(k)} = q_k \cdot \sup CT_{m|k}$$

$$CT_{avg} = \frac{\sum_k T_{tot(k)}}{\sum_k q_k}$$

$$CT_{avg} < TT_{avg}$$

(d3) Compute average work content per piece WC_{avg} by calculating first the work content WC_k per product with

$$WC_k = q_k \cdot \sum_m CT_{m|k}$$

$$WC_{avg} = \frac{\sum_k WC_k}{\sum_k q_k}$$

(d4) Compute number of required workers (headcount), HC of the cell (or a line) to produce the average work content by satisfying average takt time is

$$HC_{cell} = \frac{WC_{avg}}{TT_{avg}} \tag{5.21}$$

(d5) The number of workers have to be organized considering Fig. 5.9

In Fig. 5.9 is shown when parallelization (integrated organization principle) has to be preferred over sequentialization (split WC organization principle). It depends from the takt rate, the variability, and the work content of the arriving job; sequentialization is preferred for high performance production line where the "Kata" is important. The parallelization is preferred for slow takted variable tasks, i.e. variable content, which favors the integration of work to be assigned to the same person; indeed, the splitting of the task may cause coordination issues. An U-cell layout is preferred when several steps need mutual information exchange or the CT of several operations need to be balanced by operators.

For fast takted high performance cells, standardized work concepts are of utmost importance; special techniques are used such as MTM (methods time and measurements). For sequentialization the concept of repeatability is key, where reproduceability is important for parallel organizations. The sequential line has a handicap compared to the U-cell. Indeed, U-cells allow operators to perform non-consecutive operations to balance work of operators (fixed zone balancing), which is not possible in straight line if the operations to be balanced are not adjacent. On the other hand, for sequential line a technique of dynamic balancing is possible to level variability of operator's repeatability by continuing the work downstream until the handover can be executed, especially useful for low staffed cells. The circuit is easy to implement for high variability in demand; it is sufficient to add a further worker to the cell to adapt the ER; however the number of workers for the circuit is limited due to available space in the cell.

The aim is to implement a SPF transfer (eventually "make-n, move-n" if the number of die cavity is n or the logistic bin is n), but pay attention not to implement a "fake flow" by believing putting equipment as nearest as possible to each other to have a flow. To balance the operator work content, a time-operator chart such as in Fig. 4.4 may help and is in any case recommendable.

5.4 Reduced Vulnerability of Mixed-Product Cells

Generally, in the past, a different philosophy of cell design has been distinguishable between European and Japanese manufacturing operations. European tend to conceive rather single product cells (1-to-1 allocation), in order to avoid die change-over which reduces available production time. Japanese tend to implement mixed-product cells (k-to-1 allocation), prior having reduced change-over time with SMED technique. This paradigm can be up-scaled to entire automotive manufacturing plants, European preferring dedicated plants, generally. What is better?

In a mono-product plant, variability in demand is repercuted directly to the average of the output, which shows the same variation. In two single model-dedicated mono-product plants therefore, one may suffer under utilization and the other may need additional shifts. In mixed-product cells (or plants), due to the Central Limit Theorem of Statistics, the variation of the output mean shows a reduced variance given by the square root of the number of different models build. This means that a mixed-product cell (and even more an entire plant) is less vulnerable against variability of demand and the utilization suffers less a downswing of a single model, being buffered by the other models. We can therefore enounce the following

Theorem of Vulnerability of Mono-Product Cell (or Sensitivity Reduction Theorem)
A mixed-product cell is less vulnerable regarding load variation compared to dedicated mono-product cells in the case of demand variability. The variability is indirect proportional to the square root of the number of products k manufactured in the cell.

Lemma to the Theorem of Vulnerability of Mono-Product Cell (Lemma of Redundancy)
Having two identical mixed-product cells, these can compensate just-in-case downtime of one cell with the other identical cell, increasing operational reliability to supply all products of the mix; this is especially opportune in the case of production reliability has not yet been assured with total productive maintenance (TPM).

References and Selected Readings

1. George, M.O.: The Lean Six Sigma Guide to Doing More with Less. John Wiley, New Jersey (2010)
2. Ohno, T.: Toyota Production System—Beyond Large Scale Production. Productivity Press, New York (1988)
3. Womack, J.P., Jones, D.T.: Lean Thinking. Free Press, New York (2003)
4. Liker, J.K.: The Toyota Way, 14 Management Principles from the World's Greatest Manufacturer. McGraw-Hill, New York (2004)
5. Rüttimann, B.G.: Discourse about linear programming and lean manufacturing: two different approaches with a similar, converging rational. J. Serv. Sci. Manag. **8**, 85–91 (2015)
6. Tolio, T., Terkaj, W., Valente, A.: Design of Flexible Production Systems—Methodologies and Tools. Springer Verlag, Berlin (2009)
7. Rother, M., Harris, R.: Creating Continuous Flow. LEI, Cambridge (2001)
8. Inspire AG: Lean Six Sigma Black Belt Curriculum. Inspire Academy (2014)

Chapter 6
Linking Manufacturing Cells

In Chap. 5 we have developed the concept of multi-product, also called mixed-product, manufacturing cell, to maximize cell utilization and to increase at the same time the flexibility of a self-controlled manufacturing unit enouncing the following theorems:

- Theorem of Generalized Throughput
- Theorem of Lean Batch Sizing
- Theorem of Cell Product Congruency
- Theorem of Vulnerability of Mono-product Cell.

In the following sections we will develop the concepts of the third level of manufacturing complexity according to the TPS model of Fig. 2.3 dealing with linking different manufacturing cells, i.e. a complex multi-cell manufacturing environment or conveying the components of different asynchronous cells to a main manufacturing transfer line. We will see the concept of a supermarket linking the decoupled supply to demand as well as the concept of synchronous and asynchronous manufacturing lines and how a milk-run optimizes the utilization of a centralized supermarket. We will explain the paradigm change from queuing theory dominated push systems to downstream controlled replenishment theory of Lean pull systems with JIT philosophy.

© Springer International Publishing AG 2018

B.G. Rüttimann, *Lean Compendium*, DOI 10.1007/978-3-319-58601-4_6

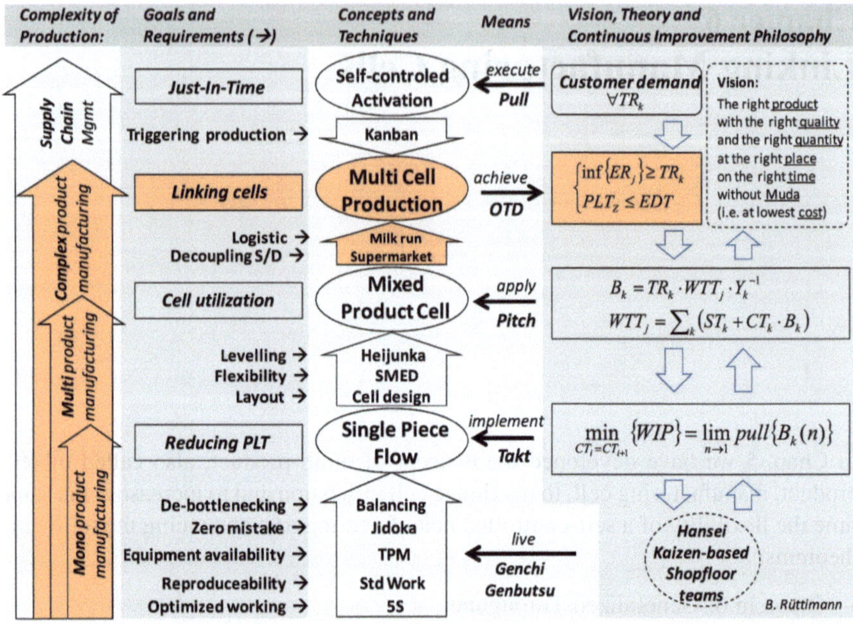

6.1 The Paradigm Change: From Push to Pull

Independent of the production principle, i.e. make-to-stock or make-to-order, Western production philosophy has been characterized mainly by the push manufacturing principle governed by complex computer-assisted MRP and later ERP systems; the consequence is to push batches through different workstations according to the scheduled routing. The formation of WIP, i.e. queued orders between the workstations according to equation system (3.7), is the consequence. The WIP between the workstations usually is not controlled and depends on the cycle times of the operations. This becomes the potential playground of Lean improvement or remains the typical application domain of queuing theory. Before performing push-pull manufacturing principles and e.g. EDD scheduling or similar principles, FIFO scheduling principle with B&Q transfer principle has been the reality in western manufacturing plants.

In the Japanese production philosophy, namely the TPS, manufacturing cells are linked via supermarkets as soon as the cycle times of the cells are different. A supermarket can also be seen as a kind of WIP (if it is not the FGI) but the origin is completely different and it fulfills a functional task of interfacing demand and supply, i.e. to separate the supply loop from the demand, two loops with different governing logics.

The control of both manufacturing principles is shown in Fig. 6.1 with the help of hydraulic modeling already introduced with Fig. 2.4 showing exemplarily the control of a water reservoir. The push principle has no explicit control once the

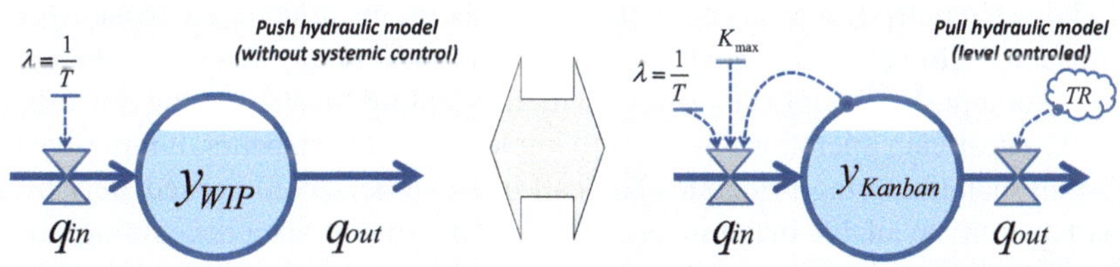

Fig. 6.1 Comparison of Push and Pull manufacturing principles with the help of the hydraulic model

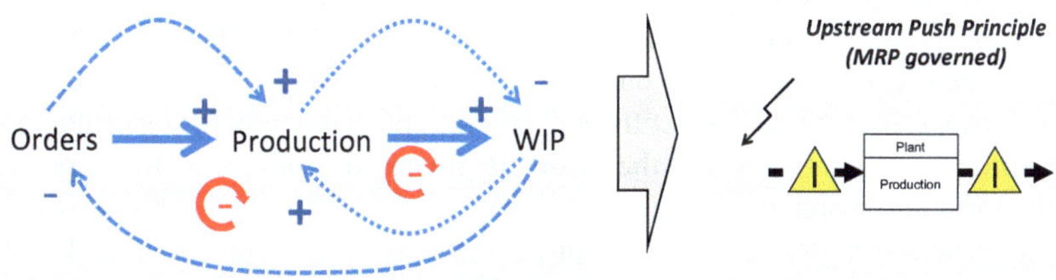

Fig. 6.2 Push principle with sporadic non-systemic control by upstream (primary approach "make-all-what-you-can")

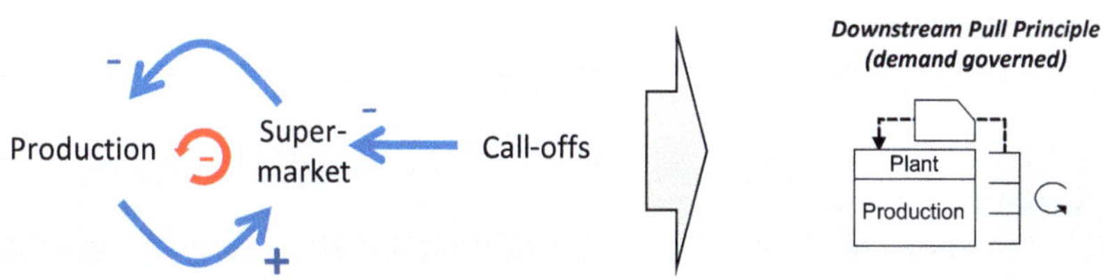

Fig. 6.3 Pull principle controlled by downstream (reactive approach "make-only-what-is-needed")

production is launched; the control of course exists, but it is only on the ERP level of production planning, i.e. a contingent managerial one-off decision. The simple causal model and the relative representation with VSM symbols is shown in Fig. 6.2. The left side shows two possible interventions: the more WIP is forming the more the production scheduling manager tries to increase production to reduce the WIP; investments might be the consequences, or the planning department will reduce order release to the shopfloor, but with the consequence that the order backlog will grow. The contrary is the pull principle shown in Fig. 6.3 which represents a self-controlled unit.

The self-controlled production-cell unit follows in principle a make-to-stock philosophy. However, the stock is not a just-in-case stock created by the consequence of a push principle but a stock with decoupling function of the downstream from the upstream operation. It serves to implement, or even better, it represents the JIT pull manufacturing principle. The call-offs can be served immediately by the small quantity available in the supermarket and the supermarket decoupling stock is replenished as soon as the Kanban (Japanese for signal) is given by the system's status itself. On the other hand, with the push principle orders have to be scheduled and represent the backlog with backlog waiting time BWT. The constant T of $\lambda = 1/T$ in Fig. 6.1 is the "mean adaptation time", which inverse value in production represents the capacity, i.e. the exit rate ER where T is the cycle time CT. The y level is controlled by the K_{max} setting which represents the maximum level of the supermarket. How to calculate the K_{max} will be shown in Chap. 7. We can summarize this section by saying: pull manufacturing principles are demand triggered and therefore takt rate TR-controlled, whereas push manufacturing principles are scheduling input rate and are therefore IR-governed. These are two opposite manufacturing philosophies.

In the nineteen-eighties, the conceptual difference between push and pull manufacturing principle was hardly understood in Western enterprises leading to incomprehension and has been provoking a still ongoing debate not only in production planning departments, if ever. Indeed, Western production systems of SME are still largely dominated by a push manufacturing principle, being pull principle against Western natural production logic. This has been reflected in the optimization of MRP systems, development, which has been leading finally to today's ERP systems.

6.2 Supermarkets

The concept of supermarkets and their replenishment are not a Toyota invention, but have been first implemented in America during the nineteen-forties. It is told, Taiichi Ohno had been inspired by this demand pull governed replenishment of commercial supermarkets and transferred the concept to its automotive industry. Supermarkets have to be understood as the "availability from shelf" of a large variety but of limited quantity of products which are replenished as soon as they are taken out of the supermarket.

Contrary to retail supermarkets, industry supermarkets are usually installed for high runner products, where the availability has to be immediate but cannot be produced to the consumption speed requested by the customer (see also Sect. 7.1 and Fig. 7.1). Supermarkets are implemented to avoid the consequence of classic ERP planning systems with deterministic, i.e. scheduled queuing systems. Supermarkets help to install self-controlled systems, which in fact are manufacturing cells with pitch-scheduled replenishment (i.e. in small batches) according to a random demand, i.e. either stochastically or deterministically governed call-offs.

Let us calculate exemplarily the dynamic behavior of the supermarket stock level according to the hydraulic model of Fig. 2.4, but now applied to the system shown on the right side of Fig. 6.1 with a Kanban controlled level of $K_{max} = y_K$ and translated to causal system dynamics as well as VSM notation of Fig. 6.3. The differential equation of first order describing the system is

$$
\begin{aligned}
\frac{dy}{dt} &= q_{in} - q_{out} = q(y) - q(t) \\
\dot{y} &= \lambda \cdot (y_K - y) - q(t) \\
\dot{y} \cdot e^{\lambda t} + \lambda y \cdot e^{\lambda t} &= \lambda y_K \cdot e^{\lambda t} - q(t) \cdot e^{\lambda t} = \frac{d}{dt}(y \cdot e^{\lambda t}) \\
y \cdot e^{\lambda t} - y_0 &= \lambda y_k \int_0^t e^{\lambda \tau} d\tau - \int_0^t q(\tau) e^{\lambda \tau} d\tau \\
y \cdot e^{\lambda t} - y_0 &= \lambda y_k \left[\frac{1}{\lambda} e^{\lambda \tau}\right]_0^t - \int_0^t q(\tau) e^{\lambda \tau} d\tau \\
y \cdot e^{\lambda t} - y_0 &= y_k (e^{\lambda t} - 1) - \int_0^t q(\tau) e^{\lambda \tau} d\tau \\
y - y_0 \cdot e^{-\lambda t} &= y_k (e^{\lambda t} - 1) \cdot e^{-\lambda t} - e^{-\lambda t} \cdot \int_0^t q(\tau) e^{\lambda \tau} d\tau \\
y(t) &= y_0 e^{-\lambda t} + y_K (1 - e^{-\lambda t}) - e^{-\lambda t} \int_0^t q(\tau) e^{\lambda \tau} d\tau
\end{aligned}
\tag{6.1}
$$

and after transforming when we set the expression

$$y_K - e^{-\lambda t}(y_K - y_0) = C$$

we observe the expected results that

$$\lim_{t \to 0^+} C = y_0$$
$$\lim_{t \to \infty} C = y_K$$

Now according to the time-depending outflow function $q(\tau)$ of Eq. (6.1) there exist different cases. E.g. for $q(t) = q_0$ i.e. a constant outflow, where in our case $q_0 < y_K$ we can set

$$q(t, t_0) = q_0 \cdot F(t - t_0)$$

and defining the step function $F(t - t_0)$ for $t > 0$ which assumes the values for

$$t - t_0 < 0 \longrightarrow F(t - t_0) = 0$$
$$t - t_0 \geq 0 \longrightarrow F(t - t_0) = 1$$

Equation (6.1) becomes better readable

$$\begin{aligned}
y &= C - e^{-\lambda t} \cdot q_0 \int_0^t F(\tau - t_0) \cdot e^{\lambda \tau} d\tau \\
y &= C - e^{-\lambda t} q_0 F(t - t_0) \cdot \frac{1}{\lambda} \left(e^{\lambda t} - e^{\lambda t_0}\right) \\
y &= C - \frac{q_0}{\lambda} F(t - t_0) \cdot \left(1 - e^{\lambda (t_0 - t)}\right) \\
y &= C - \frac{q_0}{\lambda} F(t - t_0) \cdot \left(1 - \frac{1}{e^{\lambda (t - t_0)}}\right) \\
\lim_{t \to \infty} y &= y_K - \frac{q_0}{\lambda}
\end{aligned} \tag{6.2}$$

Equation (6.2) represents the stable level of the supermarket, i.e. the maximum level of the supermarket has to be sized greater than q_0T. In reality, the outflow will not have a continuous nature. Therefore, let us try to model the more realistic case for a production environment with a time-discrete outflow.

Again, to do so, we will take the most common case for a discrete call-off type of demand in which q(τ) is a succession of impulsive functions q(t) of Diraq's δ-function applied for $t = t_i$ (where i=0, 1, ..., n), where a single impulsive function is

$$q(t) = \delta(t, t_0) \cdot Q_0$$

with $\lim\limits_{\substack{t_a \to t_0^- \\ t_b \to t_0^+}} \int_{t_a}^{t_b} \delta(\tau, t_0) \cdot d\tau = 1$

i.e. for a succession we get $q(t) = \sum_{i=0}^{n} \delta(t, t_i) \cdot Q_i$

The Equation (6.1) $y = y_0 e^{-\lambda t} + y_K \left(1 - e^{-\lambda t}\right) - e^{-\lambda t} \int_0^t q(\tau) e^{\lambda \tau} d\tau$ becomes

$$y = C - e^{-\lambda t} \cdot \sum_{i=0}^{n} \frac{Q_i \cdot F(t - t_i)}{e^{-\lambda t_i}}$$

where $F(t - t_i)$ are the unitary step functions becoming active for $t > t_i$ cumulating and remaining then constant. Please note, the replenishment is still shown as a

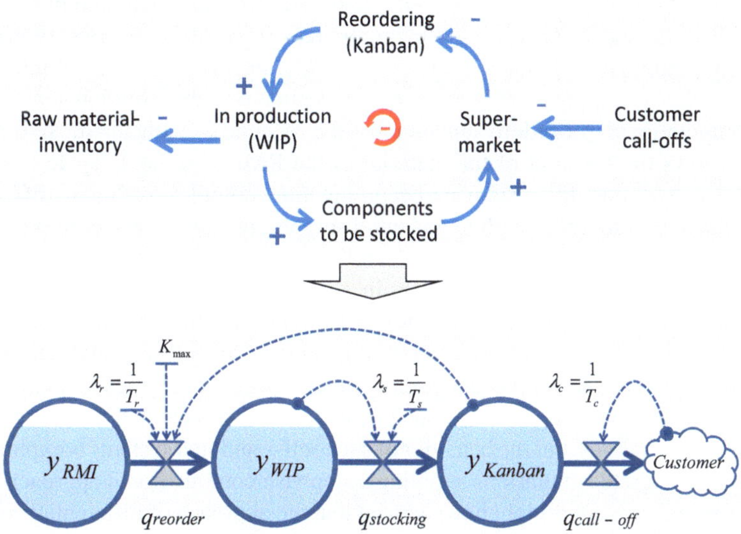

Fig. 6.4 Complete system dynamics model of a cell taking manufacturing delay into consideration

continuous function, but will most likely happen in the quantity of the pitch. For the time $t > t_n$

$$y = y_K - Y \cdot e^{-\lambda t}$$

where $Y = (y_K - y_0) - \sum_{i=0}^{n} \frac{Q_i}{e^{-\lambda t_i}}$

showing the stepwise character for every instant t_i with the quantity Q_i.

We have assumed in the left figure of 6.3, that no delay exists to put the manufactured products into stock; this corresponds of course not to reality. As we know in the meanwhile, this has to be as short as possible following the MLT_{nPF} law. Let us therefore describe a more realistic model of the system dynamics regarding our manufacturing cell which is represented in Fig. 6.4. Nota bene, in reality, for a mixed-product cell the Heijunka box introduces itself an additional BWT as small as it may be, but known in advance and buffered by the supermarket.

Manufacturing creates necessarily a WIP, which by its turn entails a delivery delay to stock; in this case it is the "white box" WIP of the usually considered "black box" manufacturing cell on the high level VSM modeling. Although the WIP might be minimal for SPF, the cycle times of the operations within the cell and the difference between the reorder and the manufacturing rate entails, that there will be a delay from the time of ordering, i.e. the Kanban signalizing to begin production, until the pieces are stocked into the Kanban supermarket represented by the MLT of the batch. This shows the importance of the MLT concept compared to the usually applied PLT; indeed, the discrete time interval ($t_{i+1} - t_i$) of integration has to be choosen appropriately to have no indesired unstable solution generated by the computational procedure and not by the system's characteristics. Here the model shows the take-out of the material of the RMI with the T_r cycle time. The equations governing this system are becoming more complex leading to differential equations of second order with exponential smoothing sinusoidal development. We will maintain our promise to limit math and refer to simulation packages e.g. [1, 2].

Due to the fact that we are ideally in a SPF regime, the reorder rate, i.e. the outflow from the raw material inventory RMI, and the stocking rate, i.e. the WIP being transferred at the end to the Kanban inventory, are proportional. Differential equations allow to calculating analytically the evolution of the stock level. In reality, due to the necessary high effort, such systems are not calculated via solving of differential equations, but the behavior of such self-regulated systems is explored via the help of discrete simulations applying a stepwise computational approach [1, 2].

The attentive reader might have noticed that the approach of differential equation aggregates the reality of single discrete events, such as a decision to replenish a Kanban-governed inventory, into a continuous succession of actions. Indeed, the difference ($y_K - y$) is evaluated in each infinitesimal instant dt, or Δt in the case of discrete simulation, and has therefore an exponential asymptotic filling approach, which in reality is not the case applying the reordering only when reaching the reorder level, following then a stepwise (according to the MLT one-batch law) or linear (according to the PLT single piece law) stock filling approach. The T_r corresponds in reality to the PIT and the T_s corresponds to the inverse of the ER of the cell and the exiting of the RMI stock has forcedly to be equal to the ER of the cell. Nevertheless, this idealization does not represent a concern when the system to be modeled is complex. Further, this does not mean, that discrete decision events for replenishment cannot be taken explicitly into account, as we have just seen for the outflow. We will not go further into this matter and refer to commercial discrete simulation packages [1, 2]. Nevertheless, we want to pinpoint, in order not to have an undesired strange behavior of the system's dynamic under evaluation, the discrete time step Δt in all from DYNAMO derived compiler types, have to be chosen appropriately, which in fact should be smaller than the smallest delay. For dynamic discrete simulation we refer to generally applicable DYNAMO-derived simulation packages such as e.g. STELLA e.g. [2] or more dedicated to manufacturing, consult e.g. [1].

Lemma of Supermarket Replenishment (or Lemma of Product Availability)

The replenishment of a supermarket from a mixed-product manufacturing cell has to be conceived in that way, that all products of the supermarket are readily available by storing the minimum sustainable K_{max} quantity, i.e. the Corollary to the Theorem of Lean Batch Sizing (Corollary of CT of a Mixed-product Workstation) applies.

In Sect. 7.3, How to seize a replenishment Kanban, we will take in consideration the delay to produce the reordered parts with a common arithmetic calculation approach to define the y_K representing the Kanban "maxloop", i.e. the number of Kanbans in the system.

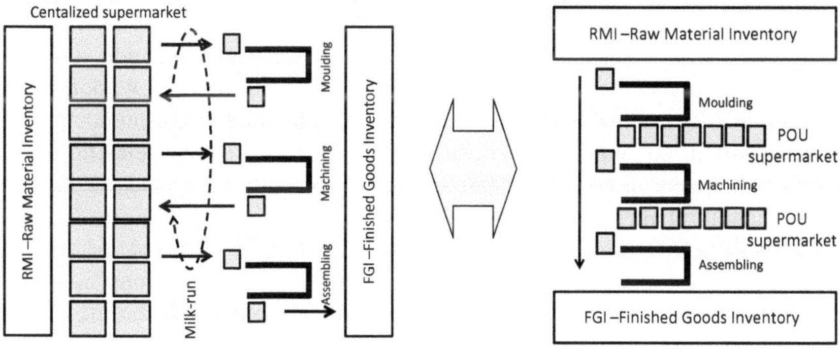

Fig. 6.5 Location principles of centralized supermarket serving POU with milk-run technique versus decentralized supermarkets located at the POU

Let us now be more practical and respond to the question where to install, i.e. to locate a supermarket. We can distinguish two main **location principles**:

- shared generic supermarket: one or more centralized supermarkets with milk-run conveying to the POU (point-of-use); the centralized supermarket needs a dichotomic Kanban system which distinguishes first a withdrawal Kanban (i.e. signaling the consumption of products from the central supermarket) and second a production Kanban (i.e. the order to replenish the central supermarket with products)
- specific POU supermarket: several decentralized supermarkets located in between the upstream and downstream adjacent cells as shown in Fig. 5.7 and Fig. 6.5, i.e. adjacent to the POU; here the Kanban device is unique but with different possible implementing techniques depending from the contingent characteristic of the operation: visual (also called triangle Kanban) usually applied for batches with technical restriction (e.g. large 30 ton casting units) and patterned scheduling (i.e. rolling aluminium coils from wide to narrow), or cards representing products (e.g. the pack-out quantity) to be produced, usually combined with a Kanban gage (green, white, red colored grid). For practical implementation suggestions consult e.g. [3].

The decentralized supermarkets allow better to attribute and make the WIP visible and therefore to reduce it further. In addition, time wasted for storing and taking-out the material has to be accounted. Of course, the supply of standard consumption parts at the POU may be supplied from a centralized supermarket with the milk-run. The centralized supermarket and associated milk-run can be implemented via AGV (automated guided vehicles); this eliminates the Muda associated to the operator distributing the required products and collecting the Kanban cards at the Kanban post. In this cases also random managed and computer controlled storing is possible and makes this option viable (we will not deepen here the topic of different central supermarket **storage principles**, such as e.g. random storage, affinity storage, frequent use storage principles). The AGV distribution of raw material and components, as well as small consumables to the POU, makes sense and has not to be confounded with the Industry 4.0 intention to let the products, e.g. transported cars, address autonomously distributed working cells; the AGV-distributed material can be performed in parallel to the production, the cars transported via AGV will generate waste in transportation, delaying PLT, a cardinal conception mistake in the case of high performance manufacturing lines [4]. For practical implementation consult e.g. [5].

Which location principle is finally applied, i.e. centralized (shared) or decentralized (specific to POU), is subject to contingent considerations; no general rule can be enounced. Pros and cons have to be established according to cost, i.e. seize/space or logistics/traffic, or simple convenience criteria.

6.3 Synchronous and Asynchronous Lines

To assemble a finished or intermediate product following the assembly scheme from different components according to the BOM (bill of material), different approaches are possible. In western production plants the predominant approach has been the time-synchronous assembly type. This approach is characterized by the simultaneous arrival of the different parts components to the assembly area. To assure the time-concomitant presence of all necessary parts, more or less sophisticated planning tools are used, such as CPM (critical path method), PERT (program evaluation and review technique) or even Petri net modeling. If a component experiences a delay, the assembly cannot start and the already available parts will congest the assembly area until all necessary parts are present, necessitating at the limit preemption, and a change of the scheduling priority causing delay in punctual delivery. It is obvious, that the main transfer principle applied is the push B&Q, although the assembly operation itself then might be organized within a SPF line.

On the other hand, supermarkets allow an asynchronous approach, such as depicted in Fig. 6.6, where both **assembly principles** are shown, i.e. the

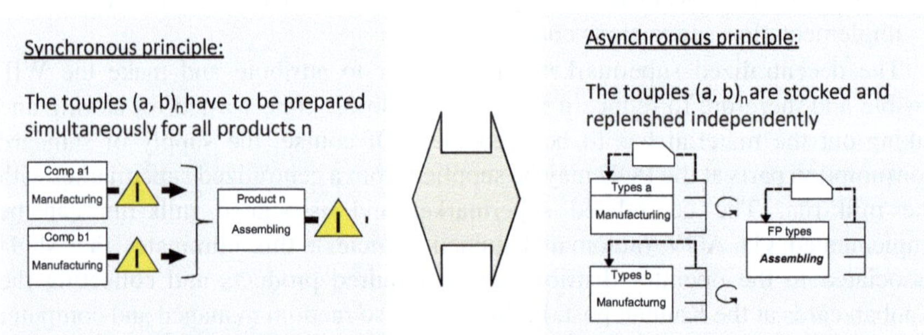

Fig. 6.6 Assembly principles of synchronous versus asynchronous assembly lines

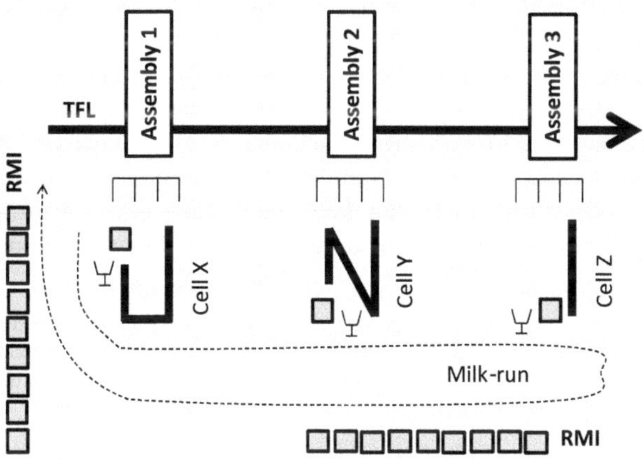

Fig. 6.7 Asynchronous level-pulled feeding of a central TFL with milk-run supplied mixed-product cells

- synchronous assembly principle, and
- asynchronous assembly principle.

In the asynchronous assembly principle the replenishment of the components supermarkets are self-controlled and based on demand-pull from the downstream assembly operation. This allows a higher flexibility of the production mix, always within a certain predefined scope of course, but is ideal to implement the JIT concept. On the other hand, mainly in Western automotive industry the just-in-sequence (JIS) is returning, which corresponds to a synchronous assembly principle, necessitating ERP systems. Toyota implemented JIS-synchronized lines feeding the final assembly TFL.

In the case of asynchronous lines, e.g. on large scale transfer lines (TFL), only the sequence of the TFL has to be scheduled (Fig. 6.7); the cells feeding the backbone assembly line will be Kanban-scheduled according to the withdrawal of the components from the small supermarket interfacing assembly. Therefore it is the backbone TFL which determines the necessary sub-component to be assembled to the product, sub-components which are ready to be extracted from the supermarkets along the line. This allows also a last minute change of the configuration of the product. The sometimes applied just-in-sequence (JIS) principle of the components, e.g. German car manufacturer BMW, sequence necessarily transmitted to the tier 1 OEM components manufacturer by the automotive manufacturer, requires a smaller buffer area, but is locked regarding the scheduling of sequence; in reality it is a step backward regarding TPS flexibility, especially in the case if a defective part is delivered. The JIS might have its reason of existence for high variability of mix, transferring the problems to the tier 1 suppliers. The JIT cells, however, are supplied by a milk-run concept, conveying the required raw material according to the withdrawal Kanban, i.e. the milk-run combines by bringing products to the POU and taking cards from the Kanban post. Note, the frequency of the milk-run needs to be calculated based on number of cells, pitch, time to get the material, and time to dispatch the material to the POU; this exercise is not carried-out in this book.

Certain rolling stock manufacturers are still assembling railway wagons from kits. Today, kit assembly operations (as well as completely knocked-down kits of automotive) are outdated for industrial scale operations. The reason for that is, that a defective part does not allow to completing to assemble the kit and therefore congesting the assembly area until the defective part has been substituted, whereas in asynchronous lines the part is put aside and exchanged by the next part available.

Lemma of Asynchronous Line (or Lemma of Maximum Flexibility)

In operations of assembling sub-components, to guarantee a high flexibility of scheduling, the supermarket's replenishment make-to-stock principle allows to maintain high performance in asynchronous assembly lines, eliminating the need for kit parts synchronization or just-in-sequence scheduling.

6.4 Requirements for JIT Manufacturing

The JIT diction is widely used but not always understood by everybody in the same way, sometimes even confounded with the limited concept of OTD to customers. Commonly, in the original sense, the JIT of TPS is intended that the product is manufactured exactly then when it is needed, i.e. JIT regarding the manufacturing of products within the operations.

JIT, as it has been interpreted by Hall (1983), is equivalent with stockless production leading to zero inventories and WIP, and therefore no delay with queuing. Edwards (1983) described the necessary requirements with the seven zeros: zero defects, zero excess lot size, zero set-up, zero break down, zero handling, zero lead time, zero surging [6].

We can synthesize all the statements by generalization, comprising the customer specific OTD, to the supply-demand relation within a manufacturing line. We can even go further and define now JIT mathematically based on the previously enounced theorems of

- SPF dominance (the transfer principle dimension)
- delay or time-trap (the minimum WIP dimension)
- bottleneck (the capacity dimension)

by defining the necessary and sufficient JIT conditions with the equation system (6.3)

$$JIT := \begin{cases} pull(SPF) = \lim_{n \to 1} B_k(n) \\ \min PLT = \min_{CT_i = CT_{i+1}} WIP \\ \sup CT_i = CT_b \leq TT \end{cases} \quad (6.3a - 6.3c)$$

or more exactly with Eq. (6.4), because Eqs. (6.3b) and (6.3c) are the conditional requirements, with Eq. (6.3a) being rather the definition of JIT, i.e.

$$JIT := \left\{ pull(SPF) = \lim_{n \to 1} B_k(n) \middle| \min PLT = \min_{CT_i = CT_{i+1}} WIP, \sup CT_i = CT_b \leq TT \right\} \quad (6.4)$$

Different than often postulated and falsely believed, Eq. (6.3a) shows unequivocally that Lean has not the aim to attain batch size $B_k = 1$, but the transfer unit $n = 1$ defining a SPF, the batch size being a consequence of the product interval time PIT and TR.

Now, the necessary and sufficient requirements for a JIT-production according to TPS at the most aggregated level, which at the end means also OTD, where OTD can be seen as the special case of JIT philosophy at the end of the supply chain, JIT can be formalized and defined with the compact form as

$$JIT := \min_{CT_i = CT_{i+1}} \{WIP\} = \lim_{n \to 1} pull\{B_k(n)\} \quad (6.5)$$

Equation (6.5) shows that minimizing WIP, which minimizes PLT, and minimal WIP is conditional to a balanced line which is attained by a pull system of a batch with the handled quantity tending to one, i.e. SPF, means supplying JIT. Equation (6.5) can therefore be enounced as the JIT necessary, but not sufficient, requirement for a stockless production leading to the cardinal Lean Theorem:

Cardinal Theorem of Lean (or JIT Theorem)

Necessary but not sufficient requirement for a JIT production, i.e. the time dimension to implement the Lean vision of the right product, at the right place, at the right time, is to strive for batch size one, intended as transfer unit, i.e. a SPF, to minimize WIP and aiming to have a balanced line.

We prefer to write pull instead of push in Eq. (6.5) to stress that the triggering is originated downward, although the difference of push and pull is vanishing according to the CLTM and Eq. (4.21) when the transfer unit tends to one. Please note, that our shortcut description of Lean given at the end of Sect. 2.2 "Lean is a Kaizen-based JIT production" is in-line with the Cardinal Theorem of Lean. By the way, this definition reflects the dichotomic character of Lean, being on the one hand a continuous improvement philosophy following Deming's PDCA cycle and on the other hand a best performing waste-less production theory.

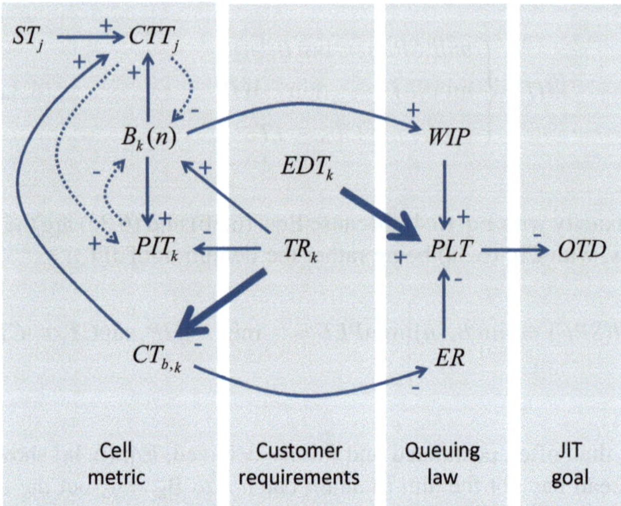

Fig. 6.8 Systemic JIT relations between TR, $B_k(n)$, CT_b, OTD

6.5 The Central Importance of TR

In Chap. 3 we learned the necessary and sufficient conditions for OTD, in Chap. 4 we learned about queuing law and lead time calculation, and in Chap. 5 all about optimal Lean batch sizing. Now let us have a look at how the synthesized math of a manufacturing cell works in order to satisfy OTD. Figure 6.8 summarizes the previous sections and shows the systemic relations between the main variables of a manufacturing system to observe OTD. Please note, as we know now, not PLT but MLT has to be confronted to EDT to match the OTD Theorem, embracing the General Production Requirements.

Figure 6.8 shows the central importance of TR in a Lean manufacturing system to implement JIT. It shows the influence it has on the key variables batch size $B_k(n)$ as well as cycle time at the bottleneck CT_b of the manufacturing cell. These two variables then influence directly the key variables WIP and ER of the queuing law determining PLT. Further it shows how $B_k(n)$ and CT_b determine the CTT of a mixed-product cell in order that the product mix can be delivered on time and how PIT, i.e. the frequency, when product k is manufactured again, is mutually influenced by $B_k(n)$. The goal is to orient every operation within the manufacturing system to the TR in order to manufacture JIT throughout the in-house value-chain. The TR is therefore the sole alignment information necessary to control the whole production, to which every self-directed cell performance needs to be oriented. Of course, due to technological constraints this orientation is not always possible; in any case the CT_b has to be faster than the required average customer TT. Nevertheless, Fig. 6.8 also reveals that PLT cannot be influenced directly but only via the driving variables WIP and ER to satisfy the customer requested EDT.

But it also reveals the central attention one has to put on the bottleneck, which can directly be influenced acting on the bottleneck.

Theorem of Lean Production Control (or Takt Rate Synchronization Theorem)
Necessary condition to have a smooth running JIT production is to align every assembly line but also the asynchronous manufacturing cells to the centrally imposed takt rate TR to implement a customer-pull JIT supply regime. The TR becomes the controlling element of the production.

Corollary to the Theorem of Lean Production Control (Corollary of Self-controlled Units)
If the OTD Theorem is not satisfied by the make-to-order principle within the value stream, Kanban controlled supermarkets (make-to-stock principle) have necessarily to interface adjacent cells, which are decentralized governed self-controlled manufacturing units, integrating production by decoupling supply and demand.

Lemma to the Theorem of Lean Production Control (Lemma of Non-prefabrication)
To comply with the Theorem of Lean Production Control, no prefabrication is necessary which would only congest the shopfloor and create Muda. Everything has to be bottleneck, or generally, pacemaker oriented.

This will lead to the Lemma of "Make to X" Production Principle which will be enounced in Chap. 7.

References and Selected Readings

1. Acél, P.: Betriebliche Simulation von Produktionsanlagen—Ein ereignisorientierter Simulator: Technomatix, Plant Simulation, ETH lecturing notes, HS2016, D-MAVT
2. Hannon, B., Ruth, M.: Modeling Dynamic Systems. Springer, New York (2001)
3. Smalley, A.: Creating Level Pull. LEI, Cambridge (2009)
4. Rüttimann, B., Stöckli, M.: Lean and industry 4.0 – Twins, partners, or contenders? A due clarification regarding the supposed clash of two production systems. JSSM. **9**, 485–500 (2016)
5. Harris, R., Harris, C., Wilson, E.: Making Materials Flow. LEI, Cambridge (2011)
6. Hopp W., Spearman M.: Factory Physics, International Edition. McGraw-Hill (2000)

Chapter 7
Triggering Production

In Chap. 6 we developed the concepts of linking different manufacturing cells to an entire production system. Supermarkets were introduced to decouple demand and supply due to different call-off and replenishment rates. We enounced the

- Cardinal Theorem of Lean (JIT Theorem)
- Theorem of Lean Production Control (TR Synchronization Theorem).

In the following sections we will deal with questions regarding the triggering of production at various levels of the internal value stream according to the TPS model of Fig. 2.3, i.e. what triggers the start of production, generally or specifically, to replenish the supermarket. When do we need supermarkets? What type of Kanban exists? How to calculate the size of Kanbans? Where to place Kanbans and especially where to install the pacemaker?

© Springer International Publishing AG 2018

B.G. Rüttimann, *Lean Compendium*, DOI 10.1007/978-3-319-58601-4_7

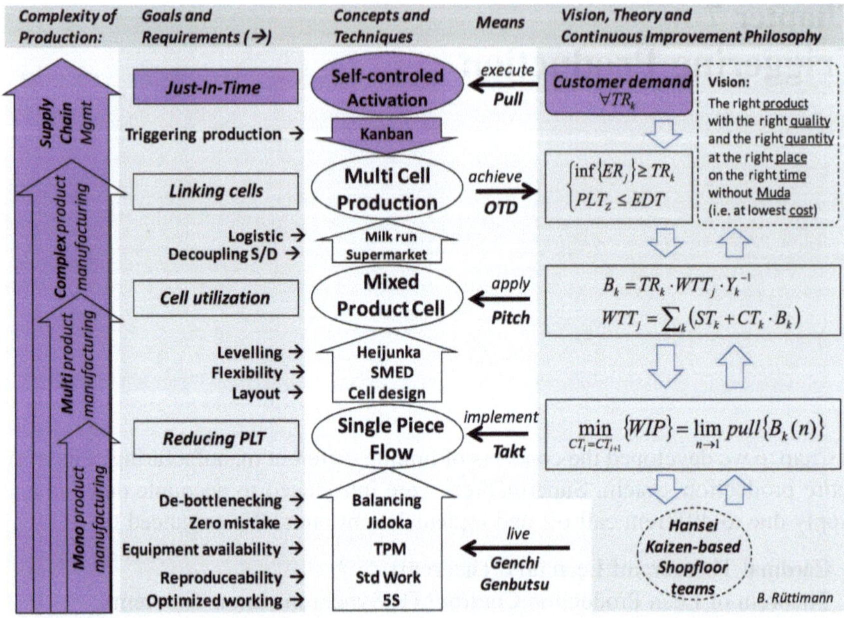

7.1 Generalized Kanban Technique

The TPS is a clearly demand-based governed triggering of production, i.e. a customer pull, implemented with the help of Kanban. Kanban is Japanese and means signal. These signals are cards, visual lamps, acoustic signs, or electronic information arriving to the computer terminal, i.e. it represents not the stock itself, but the inventory controlling device. We repeat it once again, the objective is not to install as many Kanban supermarkets as possible—there is nothing to be proud about it—but to install continuous flow whenever possible. Indeed, it may happen, that too many supermarkets are installed, which, at the end of the day, have more pieces stocked than the WIP generated by an alternative push B&Q manufacturing principle. The aim, or better the reasons of existence, of supermarkets is to decouple demand from supply.

The rule is: install Kanban when

- **the required takt rate from downstream and exit rate of upstream cell are different (cadence constraint)**
- **the production batch exceeds the demanded quantity and the product demand is repetitive (technical constraint)**
- **the expected delivery time is shorter than the manufacturing lead time (delivery requirement).**

In Fig. 7.1 is exemplarily shown the last mentioned case when an industry supermarket is needed. It is needed when the customer visible time (CVT), i.e. the manufacturing lead time MLT including backlog waiting time BWT of

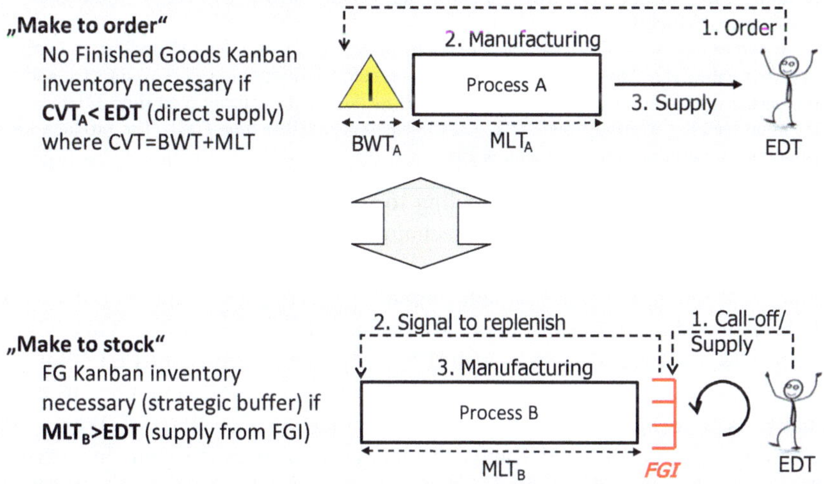

Fig. 7.1 Production principles of whether make-to-stock or make-to-order production principle apply depends from EDT and CVT

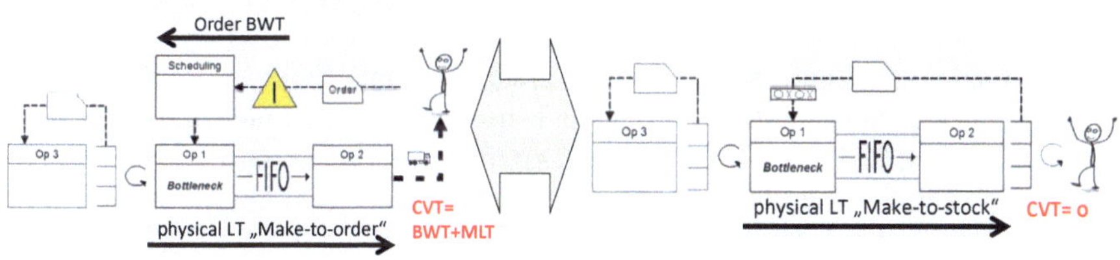

Fig. 7.2 Physical lead time and CVT may not be the same

orders is greater than the expected delivery time EDT; the same may also apply, in limited extension, also for in-house logistics by the downstream cell, which represents then an internal customer. Which production principle, i.e. make-to-order or make-to-stock should be applied is a direct consequence of the OTD Theorem. Make-to-order means here also, begin the production just when the customer needs it and the customer allows a certain delivery time slot. A retail supermarket is therefore the exaggeration of prompt availability, the supermarket of a trade intermediary comes close to the concept.

In addition, we have to distinguish the physical lead time from the customer visible lead time. The CVT comprises also the scheduling backlog waiting time BWT which may depend from the scheduling principle FIFO, EDD or with customer VIP preferential scheduling or whatever. As shown in Fig. 7.2, instead of changing the customization entry point to be able to manufacture make-to-order, which is not always possible, the production principle has to be changed into make-to-stock.

We can therefore enunciate the

Lemma to the Theorem of General Production Requirements (Lemma of "Make to X" Production Principle)

Whether the production principle "make to stock" or "make to order" applies, in order to observe the OTD Theorem, depends on the expected delivery time EDT of the customer (or on the next downstream operation) as well as the manufacturing lead time MLT and eventual backlog waiting time BWT, i.e. the customer visible time CVT. The production principle is therefore not freely choosable.

Generally, Kanban is a sign signaling to start with production which helps to limit WIP. The most general Kanban technique is "start a new order when a former order has finished" (starts equals exits, as we will see). This generalized Kanban techniques through the "inputs equals outputs" triggering-technique stabilizes WIP and therefore, due to Little's Law, a constant PLT is the result. This is very important as we have already seen in Chap. 4 having a constant PLT allows to give a predictive answer for the delivery to assure OTD. The "input equals exits" principle is a technique complying to the enounced Corollary of Weak WIP Stationarity. The generalized Kanban technique is also called Conwip (constant WIP) or generic pull—although effectively it is a controlled push. Due to the black box characteristic of the process it is irrelevant if the process follows a SPF or a B&Q principle. In Fig. 3.2 we have called that "batch on pull".

The "inputs equals exits" principle is a figurative rule of thumb as shown in Fig. 7.3; however, the management of such controlled productions, in reality, is via monitoring the WIP level. Indeed, the WIP level is measured by appropriate shopfloor data collection systems. From a practical point of view, instead of taking the output as Kanban signal, the WIP level is used to trigger a manufacturing order, i.e. as soon as the WIP level sinks below a certain reordering level. The triggering of production occurs, when the controlled WIP goes below a minimum defined WIP level otherwise wait, i.e.

$$\begin{aligned} &\text{if } WIP_{actual} < WIP_{\min} \\ &\text{then } \mathit{input} \\ &\text{else } \mathit{wait} \end{aligned}$$

The quantity of work released to production, i.e. the input, corresponds to technical restrictions and previously established batch sizes, production which may still be governed by ERP systems. Usually this corresponds to the replenishment principle of "fixed quantity at variable interval". The requirements for a

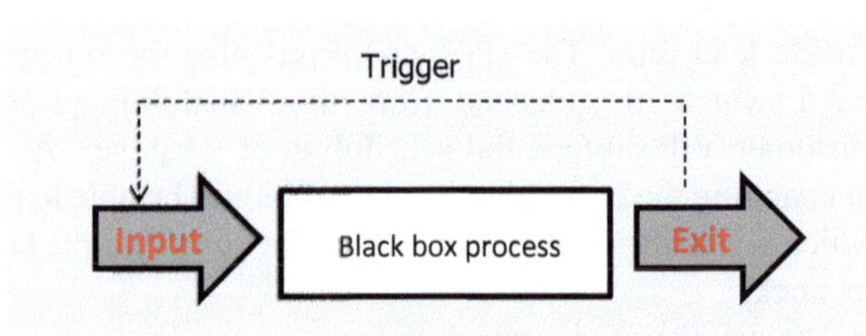

Fig. 7.3 Generalized Kanban technique with "inputs equal exits" triggering

constant WIP are given by the Corollary of Strong WIP Stationarity, i.e. in addition to the already mentioned Corollary of Weak WIP Stationarity also the Lemma of WIP Evenness must apply, leading then to a “white box” approach. But necessary and sufficient for the black box stability is the weak WIP stationarity.

Lemma to the Corollary of Weak WIP Stationarity (Lemma of “Input Equals Exit” Principle)
To observe the black box Weak Stationarity, the WIP is maintained constant with the “input equals exit” principle, i.e. without changing the transfer principle, e.g. keeping a B&Q principle, as well as in the presence of variable exit rate and without caring about the order rate, the release of production orders into shopfloor has to be equal to the exits of produced orders.

The order release to shopfloor, i.e. the input, generally and more specifically to replenish supermarkets, can follow different alternative principles. Indeed, the replenishment of supermarkets has to deal with two questions: which quantity to be replenished and at which time. To accomplish this goal, four **replenishment principles** are available:

- fixed quantity at fixed interval, or
- fixed quantity at variable interval, or
- variable quantity at fixed interval, or
- variable quantity at variable interval.

Which one is better cannot be determined unequivocally but has to suit into the replenishment tactic adopted. The fixed quantity derives from the max/min level of Kanban boards or technical equipment restrictions.

7.2 The Six Kanban Types

The classic six Kanban types are shown in Fig. 7.4. Which Kanban is ideally applied, is usually given by the problem setting, but may also be simplified by combination of two Kanbans, as we will see. These types of Kanban are:

1. The in-process (or in-line) Kanban is intended to decouple operations with different cycle times and where a pushed paced flow is not possible. This is recommended for a flow-type line where the intermediate FIFO storage has a certain capacity greater than one. Indeed, as the simulation in Sect. 4.5 showed, to protect against variability of CT the FIFO buffer necessitates a small WIP to exploit E[CT]. As we have already seen in Chap. 4, to implement the Lemma of SPF Regime, if the transfer quantity is the unit, i.e. a real SPF, it may also be possible to operate without explicit Kanban but through implicit mutual control, when the CT are not too much different (single piece handling).
2. The withdrawal Kanban serves to indicate the consumption from central supermarkets, usually accompanying a bin containing a predefined quantity of pieces. The cards are put into an intermediate Kanban-post as soon as the withdrawal

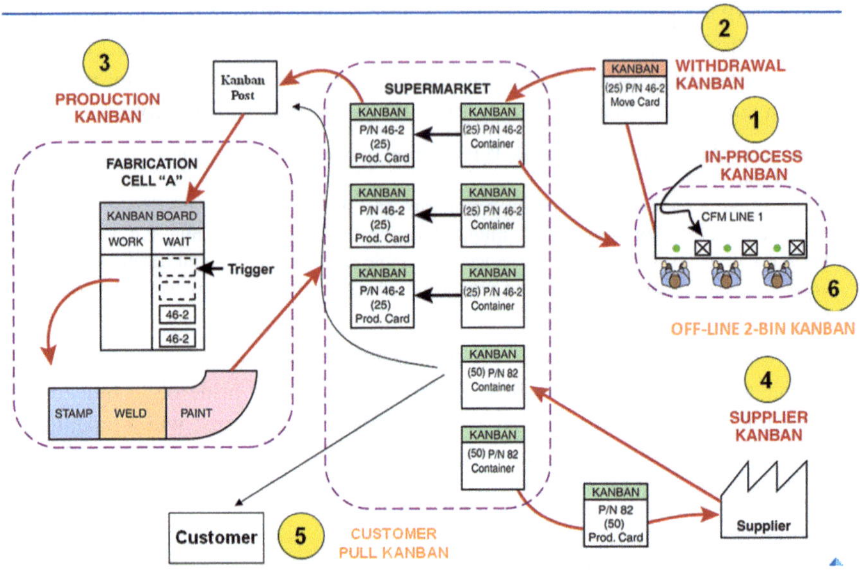

Fig. 7.4 The classic six Kanban types [from 1]

has occurred and not when the pieces are consumed; then the Kanban are collected and exchanged with

3. The production Kanban which contains manufacturing information. These Kanban cards are usually placed in a Kanban board in downward sequence, i.e. from top-down, on the production Kanban board visualizing the physical availability (Kanban stock gage with green, yellow, red zones), i.e. the balance-on-hand situation. When the cards reach the K_{min} (the green zone has been exploited and reach now the yellow zone), i.e. the reorder level, the cards are transferred "en bloc" (corresponding to the replenishment principle fixed quantity at variable interval) and put into the Heijunka box according to the pitch batchsize to allow a leveled production of several products during the same day. The Heijunka box is the cell scheduling device not needing any central ERP system, which shows the self-controlled characteristic of a manufacturing cell (within its limitation, i.e. defined product mix). Assigned to each batch is a pitch to allow that several products are manufactured the same day in order to replenish the supermarket even more than once a day (Sect. 5.1). These two joint Kanbans may be combined into one and the same Kanban card for decentralized supermarkets.
4. The supplier Kanban is the corresponding entity to the production Kanban if external suppliers are involved.
5. The customer Kanban is the equivalent of the withdrawal Kanban. Usually external suppliers are managed with Kanban cards collected at the consumption POU, i.e. shipping post and then transferred to the inbound logistics post. Transferring the Kanbans to the reception point helps to monitor which sub-components are expected from external suppliers and allows to controlling

the "on-orders". When the material arrives, it is provided with the Kanban cards and stocked into the supermarket.
6. The off-line Kanban, better known as the two-bin Kanban, is usually applied to replenish consumption material and is also comprised in the "milk-run".

The Kanban is therefore a controlling device which regulates the instant when to start production to implement JIT. JIT could also be implemented via just-in-sequence (JIS) which corresponds to synchronous manufacturing, but JIS needs a central controlling unit of production planning transmitting the sequence of products requested to the other cells (or external suppliers). The Kanban however, allows to implementing the manufacturing cells as a sort of self-sufficient units in a decentralized regulated neuronal network decoupled by supermarkets. The governing seems to be made by an "invisible hand" of production control. For practical implementation consult e.g. [2].

7.3 How to Size a Replenishment Kanban

Kanbans are usually cards accompanying products and serve therefore to limit WIP. A production is only allowed to start if the production Kanban is available signaling the need to replenish the downstream supermarket because the stock level has reached the reordering level. Therefore, any component or logistical bin containing a prefixed amount of pieces has a Kanban associated. The WIP is never higher than the maximum number of Kanbans in the loop. The number of Kanban cards depend on the activity level and process lead time, order frequency, taking into consideration that there are different products to be manufactured within the same cell. The number of Kanbans can be calculated but the real number of Kanbans needed in order to have a smooth working operation finally requires fine-tuning. It has to be stated, that the demand has not necessarily to show, but is facilitated by a fairly steady characteristics, expressed by the coefficient of variation to be smaller than one, ideally

$$c_v = \frac{s}{\bar{x}} < 0.3$$

To be clear, the demand is the OR or TR. We will use the following variables to calculate the necessary number of Kanban cards:

- the manufacturing lead time MLT, i.e. the time from production order release until parts are stocked in the inventory (which corresponds to the replenishment time). The shortest time is ideally the PLT considering when the first piece is stored, but realistically when the batch is stored being MLT_{SPF} time governed, and the longest is the $MLT_{B\&Q}$
- the product interval time (PIT) is the frequency a product is manufactured, i.e. the time interval from the last time it was produced to the new actual time it is produced again

- the demand variation buffering (DVB) stock (this is a sort of safety stock), i.e. the parts to compensate for variability in demand and supply
- the balance on hand (BoH) are the pieces physically stocked
- the on order quantity (oOQ) are the pieces in the manufacturing process and are part of the WIP.

The demand variation buffering (DVB) stock is here intentionally not called safety stock. Indeed, safety stocks are commonly used in Western European manufacturing realities and are intended as just-in-case stocks to bridge break-downs of equipment and may also not be immediately accessible. The DVB as a Japanese interpretation of safety stocks, however, are intended to satisfy an excess demand, i.e. to cover variability, with a certain service level to be defined.

Let us now put together an approximate calculation to seize the Kanban "maxloop", i.e. to seize the colored grid according to Fig. 7.5 and to comply to the Lemma of Supermarket Replenishment (Lemma of Product Availability):

$$K_{\max} = PIT_{Stock} + DVB_{Stock} + MLT_{WIP} \tag{7.1}$$

The Kanban "maxloop" is the maximum number of cards, i.e. products or logistical bins, in the production loop. The PIT component for the product k is the time needed to cover the customer demand until the product is manufactured again and is given by (see Eq. 5.4)

$$PIT_k = \frac{B_k \cdot Y_k}{TR_k}$$

where B_k is the technical batchsize which may have been calculated via the lean batch sizing formula. For a high performance cell this applies only conceptually. Note, an eventual yield Y_k which is below 100% has to be taken explicitly into consideration. The DVB component is usually given by the service level (in Western manufacturing realities these corresponds to "days of stock"!). We will simplify and say that the demand variation buffering can be put

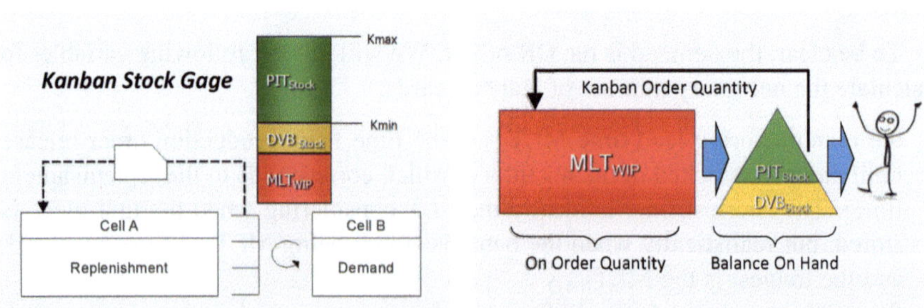

Fig. 7.5 Visual Kanban stock gage and corresponding cognitive representation of the physically available balance on hand (the demand can be a downstream cell or an external customer)

$$DVB_k = Cv_k \cdot TR_k$$

i.e. taking the TR multiplied with the coefficient of variation. This is an approximation which is fairly viable covering 84.2% of all supplies (50% + 68.3%/2). A higher service level increases the DVB stock.

The MLT component is given by the MLT_k where the MLT depends, as we have already seen, from the transfer principle SPF or B&Q but can be approximate for calculation simplifications by the PLT_k for one piece.

Equation (7.1) with the applied simplification becomes

$$\begin{aligned} K_{\max(k)} &= PIT_k \cdot TR_k + PIT_k \cdot TR_k \cdot Cv_k + PLT_k \cdot TR_k \\ K_{\max(k)} &= PIT_k \cdot TR_k(1 + Cv_k) + PLT_k \cdot TR_k \end{aligned} \tag{7.2}$$

Equation (7.2) shows explicitly that the DVB stock protects from TR variability of consumption needing a higher PIT stock. The reordering level triggering the production by transferring the Kanban cards to the Heijunka box is given by reaching the yellow zone of the gage

$$K_{\min(k)} = PIT_k \cdot TR_k \cdot Cv_k + PLT_k \cdot TR_k$$

or written with balance on hand (BoH) and on order quantity (oOQ), the triggering for transferring the Kanban cards for product k is given by

$$BoH_k + oOQ_k \leq K_{\min(k)}$$

The MLT stock, i.e. the WIP, is usually not physically available in the stock because still in the production phase, i.e. still WIP. Please note, the Kanban order quantity is not only the Kanban cards associated to the PIT, such as sometimes wrongly divulged, but comprises also the part of cards associated of the accidentally consumed parts of the DVB stock, which consumption has to be reintegrated. If the Kanban associated to the PIT are fixed, temporary Kanbans (usually marked in red) can be added. This may cause trouble within high performance cells, but will level out over a certain period of time.

The Kanban order quantity (KOQ) for the product k is given by

$$KOQ_k = K_{\max(k)} - BoH_k - oOQ_k$$

The average balance on hand for the product k BoH_k is given by

$$\overline{BoH_k} = \left(\frac{PIT_k}{2} \cdot TR_k + PIT_k \cdot TR_k \cdot Cv_k\right) = PIT_k \cdot TR_k \cdot \left(\frac{1}{2} + Cv_k\right)$$

This average BoH stored in the supermarket of the process now being integrated within a cell to supply the supermarket can be used to compare with the WIP of the B&Q principle, applied previous the Lean Transformation has been implemented. This should now be smaller. We can now summarize:

Lemma to the Cardinal Theorem of Lean (Lemma of Kanban Controlled Production)
Kanbans are used to trigger production; to control the level of WIP the number of Kanbans (maxloop) is fixed. Due to the fact that the WIP never exceeds the maxloop, the Theorem of Lead Time Stability applies, allowing JIT production.

7.4 Where to Install the Pacemaker?

Further, it has to be stressed, that the nature and the origin of WIP and supermarkets are different. WIP is the result of a push manufacturing principle, whereas supermarkets have a functional task to decouple supply from demand and represent therefore strategic buffers. WIP is the by-product of queuing production-orders in front of the machines or operations, supermarkets are the result of intentionally managed stocks to serve JIT. Because of intentionally managed stocks, the question arises where to locate intentionally these stocks. Generally, a production should have a single pacemaker, organized according the DBR concept, the "drum" being always loaded. This entails, that Kanban managed supermarkets are logically placed in front of the pacemaker. Figure 7.6 shows four different situations where a Kanban managed inventory may help to optimize production and comply to delivery requirements. The reason to decouple supply from demand is not voluntarily, but a natural consequence that supply has several technical restrictions, such as the impossibility to immediately materializing finished products, but experiencing a manufacturing lead time, but also technical restrictions due to setup and batch size. If possible, the number of supermarket Kanbans have to be kept at the minimum possible number in order to limit WIP; indeed, although we said supermarkets are not equally to WIP, nevertheless it still is work-in-process binding working capital and therefore financial liquidity.

According to Fig. 7.6, we can distinguish four **pacemaker principles**:

- pacemaker at the bottleneck
- pacemaker at the first assembly
- pacemaker at the customization
- pacemaker at the last downstream operation in a pure pull.

Independent which pacemaker principle will be applied, the location has to comply in any case to the Lemma of "Make to X" Production Principle, i.e. observe $CVT < EDT$. The first case shows the supermarket in front of the bottleneck being the pacemaker. This is the typical implementation case of the DBR principle. Based on the Theorem of Throughput, limiting the bottleneck directly the throughput and therefore the turnover and profit of the company, the bottleneck has always to be loaded, keeping downtimes to its strict minimum. It is therefore advisable to have short CT upstream of the bottleneck to replenish quickly the buffer in front of the pacemaker.

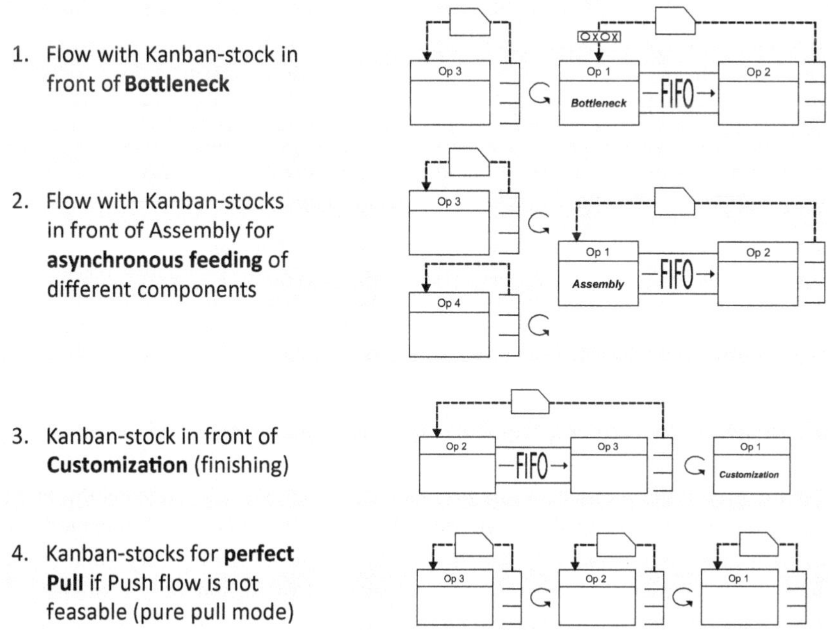

Fig. 7.6 Pacemaker principles and locations where to put Kanban stocks (leveled pull is implied) [adapted from 3]

The second case describes the already seen situation of asynchronous assembling being the assembly operation the pacemaker. This can be extended to the situation with asynchronous manufacturing cells being placed along and feeding a backbone line with different parts, such as implemented with automotive transfer lines (confront Fig. 6.7).

In the third case, the supermarket is installed in front of a finishing operation, pacemaker operation giving e.g. a customization possibility. This case allows to storing semi-finished products and adapt quickly to the changing need. One of the most well-known cases discussed in business schools during the nineties has been the approach of the Italian fashion designer and manufacturer Benetton, making its pullover not from colored thread but storing intermediate white pullovers in different sizes, needing only to be colored as a whole according to non-deterministic market demand. Generally, not only inhouse logistics, but world-wide supply chain management is becoming a strategic field of application to reduce cost and to speed-up delivery time.

The forth case shows a perfect pull manufacturing line, where between each operation—ideally a manufacturing cell—an interface supermarket is interposed. This is necessary, if the various operations, or cells, may not be linked with a unique SPF. Note, the pacemaker is always customer-pull scheduled via Heijunka-box or similar leveling concepts.

It has to be explicitly stated, that the pacemaker is always customer-pulled, responding immediately to the supplying request for a JIT and OTD. Please note, the complete linking of all manufacturing operations with a SPF, if ever possible, may not be desired, if the MLT of the SPF exceeds the EDT. Therefore, a strategic supermarket has to be put as far as possible from the end of the line, i.e. as most as possible at the begin of the manufacturing line, however, complying with the Theorem of General Production Requirements for OTD.

Lemma of Logical Supermarket Location (or Pacemaker Lemma)

The logical placement of a supermarket (independent of the physical location principle) to implement a drum-buffer-rope principle is primarily linked to the placement of the pacemaker of the process in order to assure OTD requirements or to logistics requirements such as asynchronous line-feeding or a customization operation.

Please note, consignment stocks may have the same phenotypic characteristics of supermarkets. Indeed, consignment stocks are vendor managed stocks at the customer's facility where the customer has direct access. In addition, the customer has only to pay at the moment of the take-out, the financial risks remains therefore confined to the supplier. Furthermore, how the consignment stock is managed can be different, although usually managed with reorder levels. However, whether the production and replenishment is Lean Kanban-controlled and the production is Lean Kanban-managed, is by far not a given fact.

7.5 Integrating Inbound and Outbound Logistics

The concept of customer-pull triggered manufacturing described in the sections of this chapter, extends also to the whole supply chain. Supply chain management (SCM) within a globalized economy has become of strategic importance to gain competitiveness. Usually, world-wide supply chains are designed to be cost-optimized value streams. These value-chains have experienced a huge growth also due to reduced transportation cost. Indeed, the Baltic Dry Index is a fair indicator of commercial trade and as long as it remains low, long-distance trading may be viable; low transportation cost is a necessary but not sufficient condition. Indeed, what is important is also fast replenishment. Exactly this is the mantra of Lean supply chains. For Lean supply chains the triggering and replenishment of inventories apply too, and the theory seen before can be applied to supply and delivery logistics; what might change is the name: instead of talking of PLT we will talk of supplier lead time SLT.

Apart of total cost of ownership, or total cost of fulfillment [4, 5], to which we refer, short replenishment helps to keep the stock level low, protecting against change in customer preference. We will not enter here into this topic, at least not in this first edition of this compendium, at least we want only to drive the attention to how much risk a low-cost source inherently bears in case of defective delivery. The question is: how much is a customer willing to pay for a higher priced domestic but less risky source to compensate the risk of not being supplied by a cheaper more distant but more risky source yielding higher profit due to lower cost. To calculate the individual risk premium we refer to Eq. (7.3) [6]

$$p_R = W_Z(X) \cdot \frac{D_n}{Q_n} = \sum_{x=1}^{X} \binom{n}{x} \cdot q^x \cdot (1-q)^{n-x} \cdot \frac{D_n}{Q_n} \tag{7.3}$$

where p_R is the unitary risk premium, $W_Z(X)$ is the probability of the supplier of low cost economy Z to fail X times to deliver the quantity Q_n, where q is the probability to fail a delivery and D_n is the single monetary damage caused by failed delivery. It is amazing to see how much risk a far flung supply source bears compared to a domestic source, despite higher price.

Lean extends the scope of action also to inbound logistics, i.e. how supplier of raw material and components are linked to the central manufacturing plant. Optimized routing according to milk-run concept can save money by reducing inventory. The outbound logistics has to be accorded with the customer by observing the general production requirements of OTD. Such a highly integrated macro manufacturing system has been implemented in Japan already in the late eighties. Therefore, instead of talking about TPS we should talk about a Japanese production system. Extending Lean beyond the fences of the own manufacturing plant, a synergic and optimized value-add system can be conceived where every participant will benefit becoming faster and more profitable, i.e. gaining in competitiveness. It has to be explicitly stated, that working together with non-Lean companies entails to buy and pay implicitly Muda.

References and Selected Readings

1. Inspire AG: Lean Six Sigma Black Belt Curriculum. Inspire Academy (2014)
2. Smalley, A.: Creating Level Pull. LEI, Cambridge (2009)
3. Rüttimann, B., Wegener, K.: Einführung in die Methoden von Lean Manufacturing und Six Sigma Quality Management, ETH Tools-IV Kurs. Lecturing notes HS2014, D-MAVT (2014)
4. Martinchenko, R., Grabe, K.: Building a Lean Fulfillment Stream. LEI, Cambridge (2010)
5. Jones, D., Womack, J.: Seeing the Whole Value Stream. LEI, Cambridge (2011)
6. Rüttimann, B.: Modeling Economic Globalization—A Post-Neoclassic View on Trade and Competition. MV-Wissenschaft, Münster (2007)

Chapter 8
Implementing Lean

In the previous chapters we have been going through the manufacturing theory, which stands at the base of the systemic mono-pillar Lean model of Fig. 2.3. The enounced theorems, corollaries and lemmas represent the theoretic foundation of a complex and at the same time most performant manufacturing system. Nevertheless, it is the implementation of such a Lean manufacturing system and the continuous improvement which represents the real challenge.

We will tackle only briefly the TPS philosophy and refer to the first Western publications of the TPS system [e.g. 1] or to the different variants of proliferating Lean derivatives of Lean management listed in the comprehensive overview of Netland and Powell [2]. In this chapter we will see very briefly how to implement best such a Lean manufacturing system, by following the Kaizen team-based continuous improvement approach, employing proven tools and techniques. How to organize shopfloor and what are the dos and don'ts, will deliberately not be addressed in this compendium. For this type of challenge, we refer to the many Lean guidelines and manuals available. Nevertheless, we will give in the following sections some additional hints to what to pay attention as well as some personal thoughts to the much-talked about imminent Industry 4.0 revolution, in order to comply to the subtitle "Introduction to Modern Manufacturing Theory" of this Lean Compendium.

© Springer International Publishing AG 2018

B.G. Rüttimann, *Lean Compendium*, DOI 10.1007/978-3-319-58601-4_8

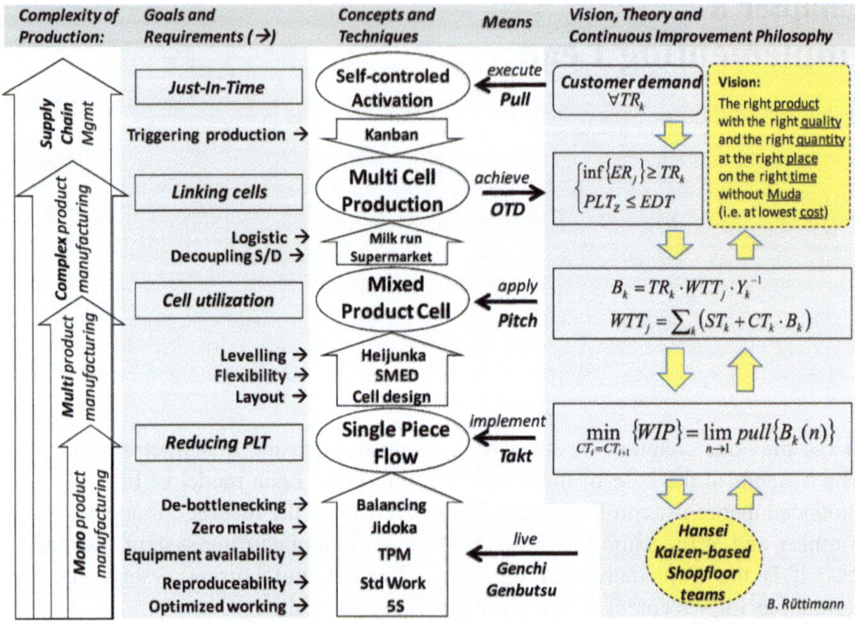

8.1 Deploying Lean and Living Kaizen

Introducing Lean manufacturing (LM) concepts in Western industries is not difficult, but needs a structured approach and full management commitment to succeed; indeed, many Western companies have begun the transformation of their production with different degrees of success but also failures [3, 4]. Important are not only highly trained employees with theory knowledge and mastering of techniques, but also management should be trained in order to capture the full potential.

Generally two dimensions have to be observed when starting the Lean journey:

- Lean deployment, i.e. the transformation of present manufacturing system (introducing Lean)
- Lean continuous improvement (living and improving the installed Lean system).

Different approaches for both dimensions can be observed, depending whether the American or European approach is followed. The American deployment approach bases substantially on the LEI proposed approach [5]. At the base of this approach stands VSM and the "present state, future state, action plan" transformation, converting value stream per value stream from MRP B&Q-push to SPF customer-pull. This is also possible to be implemented within six months via a "System Kaizen" exercise, which then corresponds rather to a Kaikaku exercise than a Kaizen. To limit the cultural shock, the European approach bases rather on organic change with Cartesian logic, beginning with shopfloor Lean awareness and 5S, visual management (Mieruka), "Point Kaizen", then introducing standardized work and so on. The two approaches begin to be confounded and mixed approaches are becoming visible. Which approach finally is applied, often depends from the selected consultancy company.

To maintain, or better to improve, a Lean system requires perseverance and leadership to instill an enterprise-wide continuous improvement culture. The PDCA application of the Deming cycle in quality has been translated into the Kaizen approach, i.e. a day by day improvement by small steps, mainly carried out by the shopfloor operators. Kaizen boards and daily meetings help to address emerging issues in daily business and to control the progress of the continuous improvement journey to attain operational excellence (OPEX). This Hansei-based Kaizen technique of daily improvement makes the operator on the shopfloor to become the central element of the transformation and improvement by giving responsibility and making him part of the success in order to achieve the vision: the right product, with the right quality, and the right quantity, at the right place, on the right time, without Muda.

Parallel to the Lean transformation another OPEX approach has been largely employed from 2000 onwards: the Lean Six Sigma DMAIC method, derived enlarging the Six Sigma DMAIC [e.g. 6, 7] quality improvement approach with LM concepts [8]. Different from pure Lean, the DMAIC approach is a problem solving methodology using Lean as well as Six Sigma tools to solve problems and improve the system e.g. [8, 9]. It is a very efficient and effective method to solve quality but also Lean issues; the cultural change towards a learning organization, however, can hardly be performed with the DMAIC method. Nevertheless, it is a much recommended method to start the Lean journey. The Six Sigma approach and Lean find their maximum effectiveness together and therefore often it is simply talked of Lean Sigma OPEX (Fig. 8.1).

Today, the concepts of LM have been translated also to the service industry with different success calling it Lean management, which embraces al functions of a company. Whereas Lean manufacturing LM tries to implement primarily a flow on customer pull, Lean management tries to minimize explicitly Muda and has generally a hidden but clear cost cutting connotation. This limited interpretation and association of Lean with Muda is perhaps one of the causes that Lean management has not yet shown the break-through in the service industries. Nevertheless, the different nature to provide a value to the customer and the different characteristic of the business makes it difficult to exploit the whole classic LM potential. A reinterpretation of LM is necessary for transactional business to maximizing the latent potential of Lean management [10].

8.2 Discovering Muda with Gemba Walk and Apply the "10.000$" Recipe

To understand a production system you cannot remain sitting in the headquarter office, you need to go to the shopfloor. Genchi Genbutsu or "going to the Gemba", i.e. going yourself to the place where it happens, is necessary in order not to be

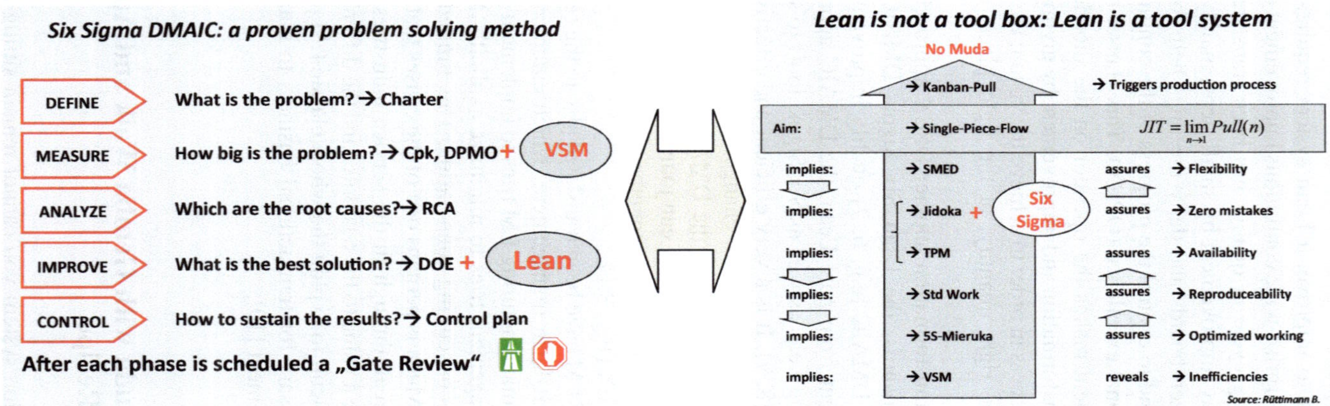

Fig. 8.1 The combined response: Within the Six Sigma DMAIC Improve phase also Lean tools today find their application. . .and so within the Lean approach together with Jidoka today also Six Sigma quality management tools are applied.

influenced by somebody else's distorted perception. While walking around, look for the classic seven wastes, or better to the nine [11] with the acronym TIMEWOODU, i.e. transport, inventory, movement of operators, ecology, waiting of products, machines, and operators, overproduction, over-engineering, defects, and unused skills and production capacity. Look around and count how many operators are working, or waiting, carrying, talking, walking, i.e. performing value-add or occupied with non-value add activities, or how many machines are running or are idled. Are safety rules observed? Are 5S-conform workplaces maintained? Are manufacturing standards respected? Are issues immediately root cause-based fixed? Is the improvement progress on track? Is the VOC (voice of the customer) known? Do you know the mood of your employees? Besides these daily routine checks, have an eye to follow the right approach in the case of an issue by setting priorities to implement fast a corrective action to strive for an efficient production system.

In the case of a bottleneck, or even worse a constraint, to improve fast an existing manufacturing system by maximizing output, the following "10.000$" recipe can be followed [11]:

- focus on the bottleneck
- create flow whenever possible
- do not fully load non-bottleneck operations (no pre-fabrication to avoid WIP)
- eliminate NVA
- reduce batch size whenever possible
- optimize cell layout
- balance cycle times of operations
- organize workplace according to 5S
- monitor WIP to achieve target (feedback for PDCA)

To manage a production system is "easy"—you only must know what and how to do it right.

8.3 Lean and the 4th Industrial Revolution

In Germany since the Hannover Fair of 2011, but already much earlier in Japan with the "e-factory" (Mitsubishi) or "smart factory" in America, the label "Industry 4.0" is divulging rapidly in Europe like an uncontrolled fire. What is vulgarly intended under the fuzzy label Industry 4.0 has a very broad scope generating during discussions more confusion than comprehension; despite that, we will not try to solve this Gordian knot in this instance but giving some aphorism with the hope not generating additional confusion, as Confucius would have told. Generally, regarding manufacturing, the concept of Industry 4.0 is to use the internet of things (IOT) by integrating it into production, forming a new cyber-physical production system (CPPS), coping demand and supply to make it fully interactive. If one reads the German guidelines [12, 13] for "Industry 4.0" (www.plattform-i40.de), regarding certain aspects, it is reading like the fairy tale "Alice in wonderland". Smart products on self-moving AGV (which already exist) address autonomously by selecting among available and convenient workstations with flexible robotized technology, integrating latest customer-desired customization options (e.g. "Porsche seat into VW" [12, 13], see Fig. 8.2); this sounds to be purposeful and desirable. Closer analysis [14, 15] reveals that maybe high flexibility and high performance were mixed-up. We have to be realistic: the Industry 4.0 revolution has not yet taken place—at present state it is more a governmental subsidized initiative leading to an uncontrolled activism. Indeed, what is lacking in this initiative is a sort of multi-generational master plan with co-ordinated tangible milestones which would give to the Industry 4.0 initiative a realistic and right sized sequence of consistent achievable objectives. This is necessary because the

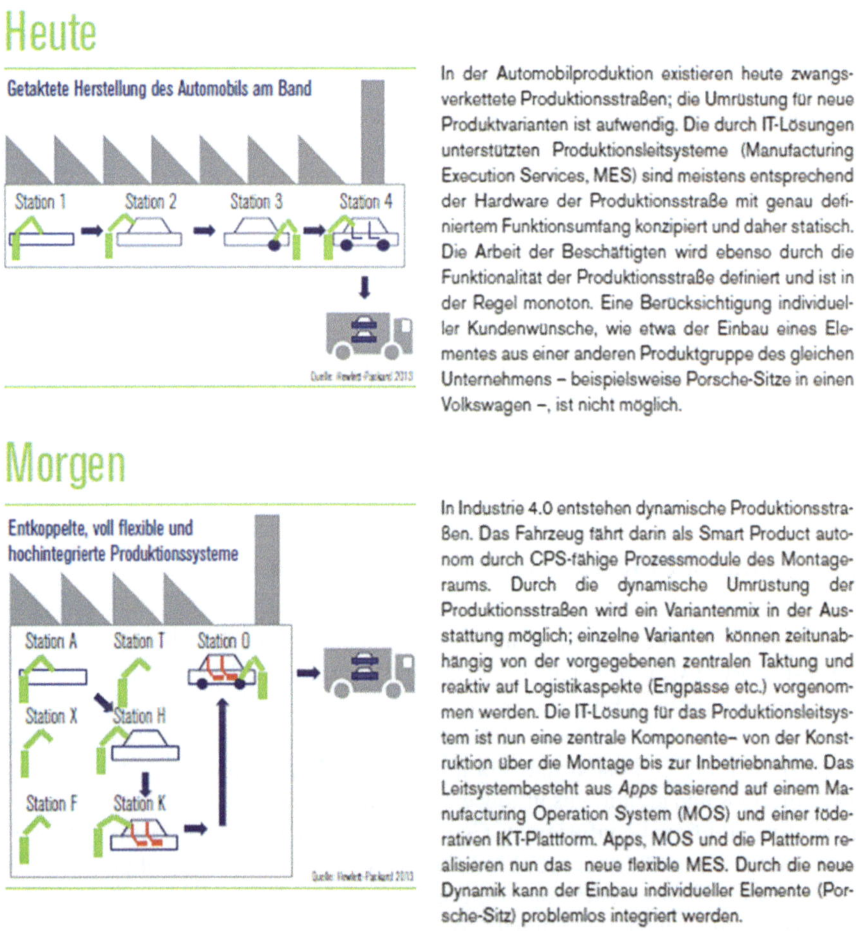

Fig. 8.2 Future state given by the recommendations of the working group Industry 4.0 [12]

complexity of Industry 4.0 compared to the 3rd industrial revolution is a multiple and it will not materialize such as "deus ex machina" in ancient Greek theatre pieces.

But now the question: How will Lean be influenced by such a vision?—For sure it will, but not as some may belief! The envisaged high automation degree of such an Industry 4.0 CPPS might stand in contradiction to the Kaizen-centric Toyota approach. However, as we have seen, Lean bears also a very solid, and above all consistent, production theory (Chaps. 5–7), based on theorems, corollaries and lemmas which, being part of a more general production theory (Chaps. 2–4), will remain also valid for flexible Industry 4.0-type of manufacturing systems. On the other hand, the high automation degree will also be absorbed by Lean-type of manufacturing systems, being automation a natural evolution of increasing technological possibilities. Further, advanced Lean companies have already reached inbound and outbound logistics, i.e. optimized SCM; IOT will only be a further technological option. From that point of view, Industry 4.0 is a self-fulfilling prophecy and the label is cleverly used by "smart" technology vendors. However, what is irritating is the envisaged high flexibility objective of Industry 4.0 for product variability ("Porsche seat into VW"). Perhaps it has to be questioned where highest flexibility is mandatory and easy implementable within a flow production. Indeed, present high performance production systems show a limited flexibility and elasticity, performance intended as fast lead time and high throughput. Flexibility means dealing with mix variability and elasticity means adaptation to demand level, i.e. changing order rate. Remember, a high but deterministic product variability, i.e. predefined mix, is also implementable with the here presented Lean techniques, as we have seen. What really irritates among the main Industry 4.0 objectives is, that automated flexibility will comprising also order quantity one, i.e. the stochastic mix variability embracing also a one-off product where the customer via IOT can interfere in the production scheduling [12, 13], at least as its final objective. If the envisaged order quantity "one" goes beyond only customization of a standard product, then this has rather the flavor of prototyping than industrial manufacturing. Indeed, whereas Lean tries to reduce variability of incoming orders as much as possible, and balancing operations as well as leveling mix and batch to implement a smooth flow without variation to maximize performance, on the other side, flexible Industry 4.0 allows deliberately to increase variability of mix and batch. This is very strange, because process performance will suffer as we have simulated in Sect. 4.5. The effects of stochastic CT and OR variability on performance. As we have seen, the consequence of variability is lost productivity and speed! Whether that consequence is intentionally and managed by purpose or rather negligence and ignorance, we will not investigate and leave it a part. Much more, the question is not Lean or Industry 4.0, also Lean will benefit from the technological innovations (see Japanese e-factory) but the domain of application, under "ceteris paribus" condition as described on the i40-platform, is for sure different: Lean TFL in the high-volume low-mix and flexible Industry 4.0 in the low-volume high-mix quadrant [14, 15]. Indeed, Industry 4.0 will rather be the solution for job shop type of problems; I did not say by purpose "for job shop manufacturing realities", which are usually SME, but "for job shop type of problems" dealing with queuing and delayed PLT, necessitating at first to prove the economic dimension and financial viability, according to the intention of the Industry 4.0 guidelines. Indeed, the necessary high investment will need throughput, i.e. a high ER, to become not only economically but financially profitable.

We will not enter here into a debate about what Industry 4.0 really might bring or not, leaving these reflections to the attentive reader. However, we will show on the next tables some comparative characteristics between classic Lean flow shops and fully flexible 4.0 systems in extremis, in order to give to the rising discussions about this topic an objective base. This is only intended to encourage critical thinking about an envisaged omnipotent cyber system which does not yet exist but which will for sure materialize stepwise in an opportune form during the next years (Fig. 8.3). We will list a comparison between Lean and Industry 4.0 of the following dimensions of system characteristics:

- transaction, i.e. the characteristic of product and demand
- control, i.e. the characteristic of systems governing
- production, i.e. the characteristics of shopfloor manufacturing system

TRANSACTION	Lean	Industry 4.0
Object	Usually: Standard product with defined customization	In extremis: Indivdual product (fully customizable)
Mix	Defined and repetitive	In extremis: undefined
Demand variability	Limited and leveled to minimize variance	Random and unleveled with high variance
Applied to:	Mainly assembly (manufacturing)	Mainly fabrication (assembly) but with broader scope

CONTROL	Lean	Industry 4.0
Objective	JIT and speed	Full flexibility
Governing pivot	Aligned to takt rate and Kanban (very simple)	ERP 4.0 CPPS considering IOT (complex algorithms) and «smart products»
Controlling concept	Decentralized self-controlled manufacturing cells via supermarkets	Central but with decentralized „intelligent" communicating workstations and products
Scheduling	Consumption-oriented decentralized Heijunka leveled pitch observing PIT	Backlog and IOT-synchronized order-batches with optimized EDD or SPT scheduling
Routing	Defined	Variable, WS load optimized
Triggering	Kanban replenishment of supermarkets	Incoming orders trigger scheduling and routing
Manufacturing principle	Pull with pacemaker	Push with interactive and dynamic optimization

Fig. 8.3 Comparative analysis of main characteristics between Lean and Industry 4.0

PRODUCTION	Lean	Industry 4.0
Production principle	Make-to-stock predominant (observation of MLT<EDT)	Make-to-order predominant (MLT<EDT ?)
Transfer principle	SPF (balanced cycle times)	SP handling (unequal CT)
Equipment dedication	Mixed-product cell/line	Isolated stand-alone flexible workstations fully addressable
Shopfloor space	minimal	fore sure more space
Assembly	asynchronous	synchronous
Bottleneck	usually defind (identified)	variabel, product dependent
Queuing	avoided	Algorithm minimized
Minimal batch	Pitch	In extremis: one
Matching S/D	Decoupled via supermarket	Integrated via IOT and optimized routing
Transport	Minimized (low Muda)	Fully flexible via AGVs (Muda)
Internal logistics	Standard milk-run to POU	Variable routing to WS
INFORMATION	**Lean**	**Industry 4.0**
Externally	Daily call-offs	Variable based on IOT, highly interactive with ERP 4.0
Coordination	Mainly visual	Electronic
Means	Kanban, 5S, Standards	IT and head-up display
Error detection	Poka yoke based	Artificial intelligence based
Out-of-control handling	RCA Kaizen-based	Self-regulating
METRICS	**Lean**	**Industry 4.0**
Lead time	Short CVT for all orders (from entry point)	Longer CVT in the average (whole line from the begin)
OTD	guaranteed	needs to be assessed
Capacity utilization	high	Low (overcapacity needed to assure OTD)
WIP within line→ WIP between WS	Low High (supermarket)	Higher Lower (optimized queuing)
Exit rate ER	Deterministic (high)	Variable (lower)
Performance	High and deterministic	Variable and lower

Fig. 8.3 (continued)

– information, i.e. the data exchange for coordination
– key metrics, i.e. the performance of the system.

The comparison of characteristics shows, that the two manufacturing philosophies follow a quite different approach. Lean is centered around self-controlling pull mechanisms, whereas Industry 4.0 will be clearly push dominated needing central production planning software, by the way, the heart of the initiative. Further, the increased flexibility might be highly paid with reduced performance and increased complexity accompanied with substantial investments in IT and redundant production capacity. Rather than reducing complexity, Industry 4.0 tries to dominate increased complexity via IT. Deducing from the Lemma of Minimizing CT Variability, now we know, that to maintain a certain throughput, variability of cycle times has to be compensated by costly overcapacity; notice, we simulated only stochastic infinitesimal variability, but not intentionally allowed discrete variability, the latter having even greater impact. The Corollary of Natural Operation Decoupling reveals that the problem will primarily not be the ER but the PLT and therefore OTD.

One of the major challenges consists in the governing pivot and controlling concept of scheduling the routing and allocation of products moved on AGV to the workstations (WS) by maintaining OTD. The mix variability increases complexity and therefore also the calculation time of scheduling and routing with algorithms of polynomial complexity. Within a dynamic changing environment and the foreseen decentralized autonomous intelligence of workstations interacting with smart products, reflecting a "k-to-m" allocation, however, this can turn the problem into a hard to solve non-polynomial problem but with exponentially increasing or even factorial-dimensional timeframe. However, the central view will be necessary at least to envisage Pareto optimality of such a multi-objective system of functions—I did not say by purpose attaining Pareto optimality—hardly being realistic. The decentralized optimization may lead to local optima which may contrast overall Pareto efficiency, but may become a compulsory way to limit the theoretic exponential dimension of the overall complexity. Due to complexity increase, neuronal algorithms may be forcedly one of the few possible ways to govern such a high flexible system. In that case the allocation, i.e. scheduling and routing, will not be the best solution derived from overall optimization algorithms but an acceptable solution as a result of decentralized heuristic approaches. Instead of applying optimization algorithms, branch and bound finite enumeration scenario simulations may be an alternative and viable approach. The non-static context and the graph structure of the problem may also induce to take dynamic programming algorithms into considerations, which are based on Bellman's optimality principle. Notice, the here enounced three cardinal production laws, the Theorem of Throughput, the Theorem of WIP, and the Theorem of General Production Requirements for OTD, as well as many others, will apply also for an Industry 4.0 manufacturing system! Therefore, whereas the technical feasibility is not questioned and will be implemented by engineers as soon as technology is reliable, issues may rather arise from planning algorithms to satisfy the customer OTD requirement. In fact, the optimal scheduling and allocation solution of the "k-to-m" problem will not come from engineers, this challenge has to be solved by mathematicians! Whether developing a new, and potentially mediocre performing push governing system compared to Lean high performance pull systems, or taking into consideration perfect and performance-exploiting, theorem-maximizing and Muda-eliminating Lean elements, has to be seriously evaluated up-front. Who knows, perhaps will Industry 4.0 become the quadrature of the circle.

On the other hand, the new IOT possibilities will also find its applications in present Lean production systems improving further an already high performance system. The flexible Industry 4.0 initiative, for the time being, is not a revolution but only a high flying initiative, which will first have to proof its performance and cost competitiveness. Nevertheless, it clearly seems to emerge that the question is not Lean or Industry 4.0, each production system, such as Lean high performance (with Industry 4.0 IOT and CPPS elements) as well as flexible 4.0 manufacturing system (with Lean theory elements), will find its suitable domain of application [14, 15]. The crystallizing question is therefore not Lean or Industry 4.0 but rather the comparison between presently low performing B&Q job shop manufacturing organizations, having also a graph structure, and flexible, potentially SP handling, Industry 4.0 systems. Indeed, for B&Q job shop production environment with less repetitive and non-deterministic defined product mix not suiting well to the Lean JIT production transformation with repetitive structure of defined order mix, high flexible Industry 4.0 philosophy may be a solution; if the high investment and reduced throughput will allow the economic viability, especially in SME, will be another question.

For sure, in the near future Industry 4.0 will materialize step by step; however, in the meanwhile we encourage you seriously not to wait for the promised land, but to start now with the more tangible Lean transformation of your company in order to shape a competitive, economically viable, proven and therefore also sustainable production system.

References and Selected Readings

1. Suzaki, K.: The New Manufacturing Challenge. Free Press, New York (1987)
2. Netland, T., Powell, D. (eds.): Routledge Companion to Lean Management. Routledge, New York (2017)
3. Rüttimann, B., Stöckli, M.T.: Reasons for operational excellence deployment failures and how to avoid them. J. Serv. Sci. Manag. **8**, 612–618 (2015)
4. Rüttimann, B.: OPEX Deployment Ansätze im Vergleich: Gibt es den idealen Ansatz? Presentation held at Lean Six Sigma Dialog Summit, November 6, ETH Zürich, organized by Gesellschaft zur Förderung der Forschung und Ausbildung in Unternehmenswissenschaften an der ETH Zürich (2015)
5. Womack, J.P., Jones, D.T.: Lean Thinking. Free Press, New York (2003)
6. Töpfer, A.: Six Sigma—Konzeption und Erfolgsbeispiele für praktizierte Null-Fehler-Qualität. Springer, Berlin (2007)
7. Eckes, G.: The Six Sigma Revolution, How GE and Others Turned Process into Profits. Wiley, Chichester (2000)
8. George, M.: Lean Six Sigma: Combining Six Sigma Quality with Lean Speed. McGraw-Hill, New York (2002)
9. Lunau, S.: Six Sigma +Lean Toolset—Executing Improvement Projects Successfully. Springer, Berlin (2008)
10. Rüttimann, B., Fischer, U.P., Stöckli, M.T.: Leveraging lean in the office: Lean office needs a novel and differentiated approach. J. Serv. Sci. Manag. **7**, 352–360 (2014)

11. Rüttimann, B., Wegener, K.: Einführung in die Methoden von Lean Manufacturing und Six Sigma Quality Management, ETH Tools-IV Kurs, Lecturing notes HS2014, D-MAVT
12. Umsetzungsempfehlungen für das Zukunftsprojekt Industrie 4.0—Abschlussbericht des Arbeitskreises Industrie 4.0. www.plattform-i40.de, April 2013
13. Zukunftsbild Industrie 4.0. www.plattform-i40.de
14. Rüttimann, B.: Von Lean zu Industrie 4.0—eine Evolution? Von einer visionären Idee zum realen Verständnis, Presentation held at: Fertigungstechnisches Kolloquium, ETH Zürich, November 26, 2015 (2015)
15. Rüttimann, B., Stöckli, M.: Lean and industry 4.0—twins, partners, or contenders? A due clarification regarding the supposed clash of two production systems. J. Serv. Sci. Manag. **9**, 485–500 (2016)

Epilogue

In this Lean compendium we have gone through the technical aspects of the TPS getting insights into the mathematically modeled logic of Lean. Nevertheless, as many believe—and we have seen the performance evidence for that—also I am personally convinced that Lean is truly the best and most performant production theory, at least presently, however, as experience has shown, it is the people involved that makes the difference in introducing Lean. To have success not only an intelligent and very knowledgeable, at the same time humble but people-motivating, management showing charismatic leadership is necessary, but also a willing, if respected, and at the same time motivation-instilled and empowered employee workforce showing engagement is needed. Only the fruitful combination of working together between managers and employees guarantees the success of a company, but also of any project, small or big it is. Although the existing destiny of being manager or employee is arbitrary, well described allegorically in Samuel Beckett's absurdist play "Waiting for Godot", the way how this mutual dependency should be lived, not passive or reactive but proactive, is at the base of real success, showing also that both characters need each other. This manager-employee cooperation is the driving force which stands at the base of any Lean transformation. This is also necessary to succeed by embracing the Lean transformation journey of a company. Nevertheless, although it is managed proactively top-down, such a transformation is not a command-act interaction, but a harmonious journey of all stakeholders of the company, requiring endeavor to succeed and to become conscious about "what's in it for me".

At the beginning of the Lean journey, most probably, managers will first go through hell, wanting never to have committed the sin of not being lean; then traversing purgatory seeing first encouraging at the same time wobbling results on the enlightening journey with small but decisive problems to be solved, but never attaining the desired heaven of Lean (also heaven maybe described by some complex mathematical equations not yet developed). A journey characterized by ups and downs of changing mood and temptation to give-up, already depicted in the fourteenth century by the poet Dante Alighieri in his comprehensive epos "The

© Springer International Publishing AG 2018

B.G. Rüttimann, *Lean Compendium*, DOI 10.1007/978-3-319-58601-4

Divine Comedy" which we could allegorically transpose to "The Divine Lean Bet". We can therefore write the last theorem of this compendium dealing with the complex hyperspace of conditional success which we try to reduce to two manageable dimensions, a last theorem with the topic we left consciously aside, but of which we remain conscious it exists, the

"Theorem" of Co-operational Success (or the Empowered People "Theorem"):
Necessary and sufficient condition to succeed with the Lean transformation, sufficient if well managed, is a harmonic interaction between managers and employees. Sharing the objectives, empowering the employees, making them a driving part of the improvement about to be proud, from executing task to solving task, is one dimension of the success' conditional hyper space. The other explicit dimension of this complex hyper space is the required leadership quality of Lean-knowledgeable managers. Leadership quality is a characteristic which has to go beyond the word management intended as pure allocation of scarce resources, but to instill and transmitting with due respect, living oneself the way to walk, the fervor to strive for operational excellence.

Although the sufficiency condition to succeed will not be itself comprehensive, and therefore might sound like an oxymoron, as well as the word harmonic may be lend from belletristic (I do not say by purpose synergic which sounds too technical in this context) Lean is by far more than a system of mathematical equations. It is also a credo, a belief in the intrinsic potential lying in the people to optimize with continuous improvement a mathematically undermined best performance production system. Indeed, a worker is not paid primarily to assemble, in extremis, this can be done by robots; a worker is paid to detect simple or complex out-of-control situations, improving the system by finding corrective and subsequent sustainable solutions to the problem assuring quality with new Poka Yoke, keeping the production always running and improving it. A worker paid primarily for problem solving and implementing improvements instead of being considered a pure workforce, which lends him again an essential role and dignified "to be" within a system of perfect orchestrated robotized working cells.

After having shown the harmonic balance of a naturally governed optimum which Lean bears, we are now coming to the end of this compendium. The last remaining act in this "divine comedy-bet" allegory is to wish you all the best on the way to reach the optimum in this endeavoring continuous improvement journey to build the most important component of which we did not talk about—nota bene, not the implementation of the equation system—but intending the heart of a company, i.e. the learning organization, for which the men's existence stands for. This told us Taiichi Ohno already many years ago; his mantra will keep its validity for sure also for the next years to come.

Overview of Manufacturing Laws and Principles

Overview of Enunciated Principles

Transfer principles
Manufacturing principles
Production principles
Scheduling principles
Main manufacturing modi operandi
Debottlenecking principles
Organization principles
Queuing principles
Allocation principles
Layout principles
Balancing principles
Location principles
Storage principles
Assembly principles
Replenishment principles
Pacemaker principles

Overview of Enunciated Production Laws

Chapter 3

Theorem of Throughput (or Bottleneck Theorem):
1st Corollary to the Theorem of Throughput (Corollary of Bottleneck Uniqueness):
2nd Corollary to the Theorem of Throughput (Corollary of Bottleneck Time-invariance):
Theorem of WIP (or Delay Theorem or Time-trap Theorem):

© Springer International Publishing AG 2018

B.G. Rüttimann, *Lean Compendium*, DOI 10.1007/978-3-319-58601-4

1st Corollary to the Theorem of WIP (Corollary of Weak WIP Stationarity or Black Box Stationarity):
Lemma to the Theorem of WIP (Lemma of WIP Evenness):
2nd Corollary to the Theorem of WIP (Corollary of Strong WIP Stationarity or White Box Stationarity):
3rd Corollary to the Theorem of WIP (Corollary of WIP Spreading):
Theorem of General Production Requirements (or OTD Theorem):

Chapter 4

Theorem of Generalized Lead Time (or Speed Theorem):
1st Corollary to the Theorem of Generalized Lead Time (Corollary of Lead Time Convergence):
2nd Corollary to the Generalized Lead Time (Corollary of Empty Line):
Lemma to the Generalized Lead Time (Lemma of SPF Desirability):
Theorem of Lead Time Stability (or Steady State Theorem):
Corollary to the Theorem of Lead Time Stability (Corollary of Strong Lead Time Stability):
3rd Corollary to the Theorem of Throughput (Corollary of Bottleneck Location):
Main Theorem of Production Time (or SPF Dominance Theorem):
Corollary to the Main Theorem of Production Time (Corollary of Lead Time Limit):
Lemma to the Main Theorem of Production Time (Lemma of SPF Regime):
Lemma to the Theorem of Generalized Lead Time (Lemma of Push Manufacturing Principle):
Lemma to the Theorem of General Production Requirements (Lemma of Flexible Scheduling Principle):
Lemma to the Theorem of Throughput (Lemma of Throughput Control):
Theorem of Debottlenecking (or Dual Solution Theorem):
1st Lemma to the Theorem of Debottlenecking (Lemma of Adaptability):
2nd Lemma to the Theorem of Debottlenecking (Lemma of High Performance):
3rd Lemma to the Theorem of Debottlenecking (Lemma of Ticketing or Queue Sharing):
4th Lemma to the Theorem of Debottlenecking (Lemma of Urgency):
Central Limit Theorem of Manufacturing CLTM (or Definition of Perfect SPF or Manufacturing Principle Identity Theorem):
Corollary to the Central Limit Theorem of Production (Corollary of Improper SPF):
Theorem of Stochastic Cycle Time Variability (or Convergent Workcontent Theorem):
Lemma to the Theorem of Stochastic Cycle Time Variability (Lemma of Minimizing CT Variability or Pull Recommendation):
Corollary to the Theorem of Stochastic Cycle Time Variability (Corollary of Natural Operation Decoupling or Push Effects):

Chapter 5

Lemma of Scheduling a Mixed-product Workstation or Cell (or Lemma of Heijunka):
Theorem of Generalized Throughput (or WTT-aggregated Bottleneck Theorem):
1st Corollary to the Theorem of Generalized Throughput (Corollary of Generalized Bottleneck Time-variance):
2nd Corollary to the Theorem of Generalized Throughput (Corollary of Frequency):
Definition of a Mixed-Product Manufacturing Cell:
Theorem of Lean Batch Sizing (or Optimal Batch Theorem):
Corollary to the Theorem of Lean Batch Sizing (Corollary of Cycle Time of a Mixed-product Workstation):
Theorem of Cell Product Congruency (or Linear Dependency Theorem):
Corollary to the Theorem of Cell Product Congruency (or Corollary of "Imperfect" Dependency):
Theorem of Vulnerability of Mono-Product Cell (or Sensitivity Reduction Theorem):
Lemma to the Theorem of Vulnerability of Mono-Product Cell (Lemma of Redundancy):

Chapter 6

Lemma of Supermarket Replenishment (or Lemma of Product Availability):
Lemma of Asynchronous Line (or Lemma of Maximum Flexibility):
Cardinal Theorem of Lean (or JIT Theorem):
Theorem of Lean Production Control (or Takt Rate Synchronization Theorem):
Corollary to the Theorem of Lean Production Control (Corollary of Self-controlled Units):
Lemma to the Theorem of Lean Production Control (Lemma of Non-prefabrication):

Chapter 7

Lemma to the Theorem of General Production Requirements (Lemma of "Make to X" Production Principle):
Lemma to the Corollary of Weak WIP Stationarity (Lemma of "Input Equals Exit" Principle):
Lemma to the Cardinal Theorem of Lean (Lemma of Kanban Controlled Production):
Lemma of Logical Supermarket Location (or Pacemaker Lemma):

MIX
Papier aus verantwortungsvollen Quellen
Paper from responsible sources
FSC® C105338

If you have any concerns about our products,
you can contact us on
ProductSafety@springernature.com

In case Publisher is established outside the EU,
the EU authorized representative is:
Springer Nature Customer Service Center GmbH
Europaplatz 3, 69115 Heidelberg, Germany

Printed by Libri Plureos GmbH
in Hamburg, Germany